Nature's Silent Music
A Rucksack Naturalist's Ireland

Philip S. Callahan

Nature's Silent Music

Copyright © 1992 by Philip S. Callahan

Acres U.S.A., Publishers
Box 9547, Kansas City, Missouri 64133-9547

ISBN: 0-911311-33-5
Library of Congress Catalog card number: 91-076335

The author and friend in Belleek, Northern Ireland, on Border (1944).

This book is dedicated with affection to Agnes and Anthony of Tawnynoran,

and to

all the McGees past and present who lived in the beautiful cottage by Keenaghan Lake.

TABLE OF CONTENTS

About The Author

Philip S. Callanhan was born August 29, 1923 in Fort Benning, Georgia. He entered the U.S. Army Air Force in 1943, where he was trained in navigational communications, and assigned to service in Ireland.

After the war, he worked in Japan rebuilding Japan's air navigation system. Later, he was in charge of maintaining radio navigation centers for Japan, Korea the Philippines and the entire South Pacific. In all, he rebuilt sixteen low frequency radio stations. Concerned about the closing of China after the war, he left Japan to hike around the world. While hiking and hitchhiking across Asia and the Mideast, he worked as a freelance writer and photographer.

Upon returning to the United States, he married Winnie McGee and started college, later earning his B.A. and M.A. from the University of Arkansas and Ph.D. from Kansas State University. He has served in research positions throughout the South and has been awarded with numerous citations for excellence in research. He is the author of some one hundred scientific papers and thirteen books. His best known works are *Tuning In To Nature; The Soul of the Ghost Moth; Ancient Mysteries, Modern Visions; A Walk in the Sun;* and texts on insects and birds. He lives and works in Gainesville, Florida and remains a world traveler.

Moreover, he has an international reputation as an entomologist and ornithologist, and has been responsible for breakthough discoveries in both areas. Most important, he is a generalist, and this—his publishers can be pardoned for saying—has accounted for insight and discoveries that arrive only once every generation or two.

AUTHOR'S NOTE

The history of the world is written in the wars and killing deeds of a nation, but that is a dead history based on the bigotry and hates of people. Living history lies in the hearts and souls of humans, and in their day to day activities. History, one might say, is a "fit" of a people to the lay of the land. If the fit is close, as with a tongue to a groove, then life can be good; if the fit is poor then the tongue and groove of life and land will slip apart and disaster will follow. Disaster usually takes the form of murderous wars and famines of which there are plenty of both at this writing.

This is the story of my forty year look at a rural people in a little spot in Ireland that fit the lay of the land very well indeed. It was my love and belief in those gentle people that led me to search for the fairy and leprechaun forces that I felt all around me. In the end it led me, as a naturalist, to rediscover the low energy forces known before only in different terms to the ancients. These forces may someday be truly understood, and thus once again utilized in agriculture and healing.

In this book you will find nothing of the fanaticism with which a few evil people on all sides torment mankind around the world, and especially in Ireland. That is the legacy not of fairies and leprechauns, but of the devil—it is the legacy about which historians so love to write and which tells us nothing about the heart of mankind. Strangely enough, it was this same evil distortion of man's spirit that so disrupted my life and landed me in Ireland for the first time.

It is also not the legacy of the world's farmers who are peaceful and productive until those of greed seek their land for power—That is what the Irish "troubles" have always been about.

The word *naturalist*, according to the dictionary, means "one versed in Natural Science." The natural sciences used to include physics, chemistry and biology but in these days of specialization the term is applied to biologists alone, and never to the farmers, whereas in reality the farmers of Ireland, and of yesteryears in the U.S. were naturalists in the truest sense. They understood nature and thus the weaknesses of any agricultural method that works against nature instead of *with* nature. Why have modern farmers been so corrupted as to slay their hedgerows and soil. I believe it was best put by Christopher Derrick in his masterpiece "The Delicate Creation" when he wrote in his chapter on the relationship between twentieth century man and his environment:

"In the second place, we can dissociate the question of this world's origin and nature from any question of divinity or supernatural power, evil or good. This planet and the entire universe will then be seen in totally non-religious terms, though lip-service may still be paid to a remote and theoretical God. The sense of creation will not be strong: things will seem to have evolved automatically, mechanically, by the blind operation of natural law, and an attitude of neutrality of indifference towards them will seem appropriate and may be psychologically inescapable."

God is dead and so is viable agriculture.

The first chapter of this book is from my original notes written by recall in narrative style shortly after World War II, some portions are also taken, with permission, from a series of articles published in *Ireland of the Welcomes* between April 1979 and the present.

Old photographs herein appear grainy and of nostalgic character because they were taken with a vintage miniature Foth Derby camera, one of the earliest pocket-sized cameras produced.

During World War II diaries and cameras were required to be sent back home from the port of embarkation. I escaped overseas with my little Foth Derby because during inspection, officers searched duffel bags and food lockers. My camera was in my pocket. It was walking in the countryside of Ireland with this camera that first brought me in contact with the ecoagriculture farmers of Ireland, and the remarkable McGee farm of Bonahill. I thank the McGees for their vision and especially my wife Winnie McGee Callahan for her love and inspiration.

FOREWORD

It is an honor to write an introduction to this lyrical book. Today, as I begin the task, the "Science" section of the *New York Times* [Tuesday, 4 June 1985] has a front-page story entitled: "Prestigious Forum Slides Into a Troubling Decline." The forum is the annual meeting of the American Association for the Advancement of Science, and the decline is the sharp dropping off of attendance at that meeting.

While, as the article makes plain, no one is quite sure what factors are at work to explain the drop-of, there is a sense that the increasing *specialization* of science is now eroding the former popularity of general meetings that focused, not just on narrow concerns, but on important scientific and public policy issues of consuming interest to a wide spectrum of scientists.

A real danger in the decline, the article stresses, as do many senior scientists, is that future generations of researchers will lose the cross-fertilizing benefits of learning what scientists in other fields are doing. They may also lose the opportunity to engage in important policy debates about the impact of their work on society, and thus edge into positions of servitude to whomever will make policy.

Whether young scientists, being spawned in ever greater numbers, even care about the social impact of their work is doubtful in light of their apparent lack of interest in applying their knowledge to problems connected to the ultimate fate of the earth.

At the close of the Orwellian year of 1984, an epochal meeting was held in Ottawa, Canada, that was characterized by the *New York Times* as "taking the first step toward what may be the biggest scientific efforts ever." Hosted by the International Council of Scientific Unions, it unanimously endorsed a resolution calling for a worldwide project: the study of the powerful, and complex, interaction of our planet's physical, chemical, and biological processes. Sir Jon Kendrew, president of the august assembly, did not mince words when he told the press: "It's really taking a look at what happens to the future of the human race."

Given both the urgency of terrestrially pressing problems and the scope of what would amount to a "rescue mission" called for by the Ottawa conference, it is all the more startling to read, at the end of the *Times* article, some truly grotesque commentaries.

The newspaper called attention to "serious reservations" on the part of many scientists who seemed to feel queasy about the project because, in their view, it "would divert funds from other more pressing work." Some of these self-styled "thinkers" went even further, insisting that, because of the sprawling scale of the endeavor, it would be difficult to recruit people to work on it because it would "last longer than the average researcher's career." To supplement that bit of nonsense it was stated that a majority of the scientific fraternity believe they are most useful when pursuing pure research in a narrow discipline rather than in "uncharted interdisciplinary fields."

Anyone who reflects upon these statements must be led to

ask a few questions, one of which would be: what on earth could be more "pressing work" than that of safeguarding earth itself? Another would be: how can anyone in his right mind consider his "career"—ostensibly meaning the limited goals of his limited life—to be more important than that of safeguarding?

These are the kinds of questions Phil Callahan would ask, as proved by the fact that, in his writing, he alludes in forceful language, to the basic, and proliferating shortcoming of science: a self-imposed limitation on the breadth of horizons considered fit for viewing.

Callahan's master opus, *Tuning In To Nature: Solar Energy, Infrared Radiation, and the Insect Communication System*, written in as sprightly and excitement-producing a style as that of the best detective thrillers, unravels for the first time such mysteries as why moths suicidally approach candle flames but not other sources of fire, why a cabbage butterfly zeroes in on a cabbage but not a potato, and why millions of "love bugs"—so-called because they copulate in mid-air—are attracted to paved highways in the southern United States, there literally to coat them with a slimy, slippery ooze to the point of creating a serious traffic hazard. In this work Callahan warns his readers to keep in mind that "despite my computer and my laser, my electron microscope and my interferometer and my radiation detectors, my way is not the way of the technologist, but rather the way of the naturalist."

Naturalism is a philosophy one of the principal tenets of which is "conformity to nature." Since a basic meaning of conformity is "compliance with usage," naturalism stresses compliance with nature's way. But for more than a century, unnatural efforts have accented, not compliance with, but domination of nature, and the results stemming from this brutal approach have led to the convening of such a conference as the one held in Ottawa.

Another criticism of Callahan's is that scientists no longer write for popular consumption. Not only does nature escape the scientific specialist who excessively relies on elaborate apparatus to examine simple natural phenomena, he says, but also the one who uses excessively elaborate or arcanely opaque language to describe them.

This is not what budding scientists are taught in graduate school these days. Indeed, to write in other than a specialist's jargon appears to be a cardinal transgression. Another violation of so-called "scientific" conduct is never to write in the first person, as if to use the word "I" implies that a mere human is at work. "This attitude," scoffs Callahan, "is inappropriate because it gives to the literature of science a facade behind which its authors appear to be infallible and is the reason why science which is, or should be, the study of nature, and thus altogether fascinating, is presented to mankind in such a dull guise."

Still a third transgression is to write science on subjects outside the field for which one is trained—as scientists sanctimoniously put it, "outside the area of one's competence." Refreshingly, Callahan has no fear of, or respect for, that "sin," as proved by his latest book, *Ancient Mysteries, Modern Visions: The Magnetic Life of Agriculture* (1984) in which he presents fascinating data on the effects of stone on living creatures and the soil, and develops a brand new theory to explain why the round towers of Ireland—long considered to be only fortresses protecting their medieval occupants from marauding bands of Vikings—were actually constructed to enhance the productivity of the soil and, thus, agricultural harvests.

Callahan admits forthrightly: "I have no formal credentials to *prove* to my readers that they should take anything that I say seriously. And in this age of overspecialization, that surely is a transgression." But, more significantly, he adds: "I present the most powerful credentials of all: those of repeatable experiments."

To understand Callahan, who refers to himself as the "philosopher" as much as the "scientist," is to be aware of some of his heroes to which he pays homage in his book, *The Soul of the Ghost Moth*, one of the world's shortest autobiographies. Its very title was inspired by South African Eugene Marais' *The Soul of the White Ant*, a now classical book on termites written nearly fifty years ago. A journalist, lawyer, and poet, Marais was—like the Belgian writer-dramatist and Nobel laureate, Maurice Maeterlinck, who wrote a book on honey bees—above all inspired by his love for the study of living creatures in their natural state.

So is Callahan, another of whose idols is Jesuit and natural philosopher, Teilhard de Chardin. His book, *The Phenomenon of Man*, Callahan characterizes as a "philosophical masterpiece," Like Teilhard, Callahan believes that organic evolution is God-directed. He also admires Teilhard for being one of the first to show that science and religion are *not* antagonists as the age of technology—or so-called "reason"—would have us believe. "They" want us to hold their belief because science is averse to including "mind" or "consciousness," far less "soul" or "spirit," in the area of its concern. What is the reason for this? Is it a disease? That is the conclusion of a scientist of stature, Sir John Eccles, neurophysiologist and Nobel Prize Laureate for Physiology and Medicine, who in his book, *The Self and the Brain*, writes:

"Another disease is the materialistic or mechanistic philosophy held by so many scientists who reject or ignore all the phenomena of conscious experience . . . I an reminded of some physical scientists who facetiously regard life as but a disease of matter and would restrict the scope of science to the inorganic world. Scientists who reject the phenomena of conscious experience from the domain of science . . . would also reject imagination, critical thought, evaluation and judgment which are so vitally concerned with conscious activities as one wrestles with scientific problems."

In his writings, Callahan epitomizes the opposite tendency He accepts what the scientists Eccles' talks reject. And not only that, he incorporates it into the fibre of his professional activity. Thus, in the present book, Callahan continues, as he did in the previous ones, inseparably to weave his life and his work into unifying whole cloth. Part of his incentive and inspiration for this came from the Irish field botanist, Robert Lloyd Praeger, whose masterpiece on the natural history and topography of Ireland begins with the words: "The way I went was an Irish way." From that opening sentence he took the title of his book, *The Way I Went*, which is both a natural history and, in Callahan's words, "a field guide to the soul of an entire country."

Callahan revels in Praeger's making no scientist's apology for the autobiographical nature of his study. "Like Praeger," he says, "the way that *I* went, and where I am today, both as a scientist and a human being, has most certainly been an exciting journey. As a researcher, I can no more separate my emotions from my science than from the rest of my life—nor can any scientist."

Another half-forgotten scientific giant admired by Callahan was the Irish biologist-physicist, John Tyndall, who wrote *Heat Considered as a Mode of Motion*, in which he described how molecules of perfume scent such as those emitted by Pachouli, sandalwood, oil of cloves, lavender, attar of roses, lemon, thyme, rosemary, oil of laurel, sikenard, and aniseed *absorb* infrared radiations. Tyndall's focus was thus on the action of scents on radiant heat.

Before writing *Tuning In To Nature*, Callahan had never run across Tyndall because he was totally ignored by modern chemists and olfaction physiologists. Despite the fact the Irishman was the father of an entire science, infrared spectrophotometry—the basis for all analytical chemistry—his contributions were characterized in the 1953 edition of the *Encyclopedia Brittanica* as "due more to his personality and his gift for making difficult thing clear than to his original researches." Nor is it widely known that he discovered the effects of penicillin eighty years before Sir Alexander Fleming.

Callahan was instantly attracted to Tyndall's findings because, approaching the same problem that the Irishman had worked on but from the opposite direction, he had been studying the action of radiant heat on scents for a quarter of a century. His tribute to his predecessor and analysis of his work in this volume is as graceful as it is incisive.

One remarkable discovery to which Callahan's years of

meticulous research led was, in simplest terms, a demonstration that organic molecules could be stimulated to emit infrared radiation by simply trapping, or vibrating, them at what are called electromagnetic Extremely Low Frequencies (ELF), or those at from 1 to 1,000 Hertz (cycles per second). Callahan consequently came to the firm conclusion that the radiations he worked with are the very control frequencies of life itself and that much disease is the result of distorted communications between organic control chemicals and cells.

"Though there is no mathematical, or theoretical, model for my discovery," he wrote in 1977, "this dos not make it any less a scientific fact."

Now, nine years later, work done by him and in other quarters has begun to establish the lacking theoretical and mathematical model and to disclose a Pandora's Box of effects from ELF magnetic field which, applied aggressively, may make of the nuclear threat a sort of atomic Maginot Line.

In May 1984, at a meeting in Atlanta, Georgia, Dr. Eldon Byrd, a biophysicist working for the U.S. Navy, unofficially revealed what highly specific Extremely Low Frequencies can now do. Among other things, they can: 1) alter the behavior of cells, tissues, organs, and whole organism; 2) alter the levels of hormones in living creatures; 3) alter the reaction time of irreversible chemical processes, as well as the chemistry in the living cell itself; 4) alter time reception in both animals and humans; 5) make animals go to sleep; 6) make bone grow or stop growing; 7) start, or stop, cell de-differentiation, meaning that a cell, programmed to one thing, can be re-programmed to do something else; 8) regulate immune processes; 9) destroy cells by rupturing them; 10) cause tremendously profound defects and alterations in embryos during their gestation periods by directing interacting fields with growing fetuses; and 11) cause up to six times higher fetus mortality and birth defects in laboratory animals with fields so weak they can barely be detected.

On the more fortunate side, other frequencies, properly selected, bid fair to offer life-enhancing, instead of life-destroying, properties.

This is the essence of Callahan's philosophy: the enhancement of life. Astutely developed and applied, his discoveries about the behavior of insects could lead to electromagnetic controls of their behavior which would completely eradicate the need to use dreaded Agent Orange and a host of other lethal herbicides and pesticides being wafted daily onto our forests, farms, and gardens. The fact that the money has not so far been allocated for such development and application speaks volumes about how interests vested in billions of dollars of profits from the manufacture of these chemicals persistently, and covertly, exert pressure against action that would relieve us of these monstrosities. In this loving and gentle book, Callahan shows how "civilized" insanity is turning Ireland away from her naturally harmonious, aesthetically pleasing, and sound practices. And in what direction? The same one which everywhere would remove a thatched roof to replace it with galvanized tin, only to increase the heating bill. The same one which would rip out miles of hedgerows only to destroy dozens of species of birds that nest in their protective thicknesses. The same one that, in the name of "development," is cementing over, or otherwise desecrating, hundreds of square miles of woods and prairies, polluting underground aquifers, ponds, lakes, and the oceans themselves, and fouling the air that we, and our fellow creatures, breathe.

Who, following in the steps of Robert Lloyd Praeger, will write "a field guide to the soul of the earth?" Maybe that will be Phil Callahan's next undertaking. He, one of the few among natural philosophers, is capable of bringing it off.

Christopher Bird
co-author of *The Secret Life of Plants*
and *Secrets of the Soil*

INTRODUCTION

It was pure fate and the "battle of the beams" that landed me in Ireland during World War II.

The story begins over forty years ago, shortly before D-Day. As a young twenty year old G.I. in what was then the Army Air Corps, I was at the port of embarkation in Sacramento, California, waiting to be shipped to the South Pacific. My Pacific orders never came; instead, I was put on a train with orders to proceed back across the United States to New York.

Although I was still in khaki, my travel orders described my secret destination in short, terse military language—cold, wet, and windy!

I had no idea what area of the globe my orders referred to. I was a highly trained radio technician with a specialty in a type of low frequency transmitter called a radio range. The early Adcock radio range transmitters produced an invisible beam, a sort of electronic highway through the sky. The beam was used by anti-submarine pilots to find their way across land and sea.

It was my job to see that the invisible beam never disappeared from the airways.

I belonged to a unique outfit called the Army Airways Communication System—AACS. AACS had outpost installations from the steaming tropics of the South Pacific to as far north as the Arctic Circle. Its main headquarters was in Asheville, North Carolina. From there AACS personnel could be assigned to anywhere on the face of the earth. The immediate result was that the first ever world-wide communication network emerged as if by accident.

In 1944 AACS radio operators could relay a message around the world almost as efficiently as do satellites today. In the few short years since it was invented by Nikola Tesla, radio has enveloped the globe as a sort of electronic nervous system. Life as man knew it would never again be the same. Intercourse between countries that once took years or months to complete was suddenly compressed into hours or minutes. Einstein was correct: Time, for all practical purposes, is relative.

If time has speeded up for the rest of the world, it was for me, despite my involvement in complex electronics, about to slow almost to a standstill. My life was, in fact, about to reverse and go backward in time. At the ripe old age of twenty I discovered that a real, functioning time machine can exist and that like all such relative matters it depends on the space that one occupies.

My fellow G.I.s at the airbase in Sacramento insisted that cold, wet and windy meant I was headed for the Arctic Circle, no real price for me, a bird watcher since age twelve, with visions of parrots, myna birds, and tropical forests.

My companions were wrong, for within three weeks, after a rough North Atlantic crossing on the great ocean liner New Amsterdam, I found myself, now dressed in wool, headed for the little Scottish port of Stranraer. That day was the beginning of my backwards journey through time. I shall never forget that early morning in Scotland when I first reverted, literally into what was another time-space frame.

At noon we docked in Ireland. I stepped ashore into the

nineteenth century!

In the United States during the early 'thirties, and prior to rural electrification, some Americans believed they lived in the nineteenth century. But they did not, in fact; the sudden modernization of America after World War I created an atmosphere completely different from the one I was about to be enveloped by. In reality, it is atmosphere that I am talking about.

Like the hero of the marvelous musical *Brigadoon,* I had wandered into a Scottish mist at the seaside town of Stranraer and emerged across the Irish Sea into a magic village called Belleek. There I was to spend two years among its delightful inhabitants.

Of course, the dwellers of Belleek did not realize that they lived in a magic village: people who live in magic places never do. What made my journey a true-life *Brigadoon* was the extreme contrast between the modern circumambience of my upbringing in middle-class America, and everyday life in Belleek in the early 1940s.

It did not take me long to realize that fate had deposited me in an evanescent countryside. It has taken me forty years, however, and a dozen trips back to understand the significance of that journey—the simple fact that it influenced me to question the modernization of the world, and in particular the agricultural depredation of Ireland. I do not mean that modernization is of itself all bad, only that we should trim away the excesses and substitute a little Irish magic for certain ill-devised technologies. We should in fact begin again to believe in leprechauns.

Philip S. Callahan
Gainesville, Florida

1

BELLEEK ON THE RIVER ERNE

AT THE OUTSKIRTS of Enniskillen, an MP motioned us to the side. We had already driven one wheel up on the sidewalk, and were squeezed between a convoy and the greystone fronts of the village shops. Coming from the opposite direction was an endless line of three-quarter ton trucks loaded with infantry men of the American Eighth Division. They were headed east for England and the battlefield in Normandy. A few days before our arrival in Ireland, the Americans and British had landed at Normandy and the last of the divisions trained in Ireland were pulling out. Soon there would be no GIs in Northern Ireland except for Air Corps personnel at Nutts Corner, the airfield on the outskirts of Belfast.

"Turn around," the MP ordered, "and get on the end of the convoy."

"What for?" I asked. "We are headed for Belleek, not Belfast." "Belleek—where's that?" he said. "Let's see your orders." I pulled the sheaf of orders from the folder and handed them

The valley of the River Erne from the "Island," Belleek.

over.

"Uh, Air Corps, huh? What's in Belleek?"

"A radio station," I replied.

"How far's Belleek? I never heard of it."

"It's on the Free State border, as far west as you can go and still be in Northern Ireland," I replied.

By then, we were tired explaining to MPs that we actually were headed in the right direction, and were not trying to make it back to another village for a fond farewell to some local Irish lass. The movement of the Eighth Division from the town of Enniskillen to the front was supposed to be a hushed-up military secret, but when the convoys and loaded trains pulled out of Enniskillen station in the early morning hours, half of the people in town, including a goodly number of unhappy colleens, had shown up to wish the GIs Godspeed.

Enniskillen is situated on the southeastern tip of one of western Ireland's most beautiful loughs. The River Erne, which drains the lough, flows to the sea through the southwestern tip of County Donegal. Most of the river lies across the border in the Irish Free State, but all of the twenty-three mile long lough lies in Northern Ireland. The road from Enniskillen to Belleek winds along the edge of the lough. It is an undulating road and soon I was engaged in reading the picturesque name places along the road.

Ireland is a country of contrast and paradox. One may drive for only a few miles and yet have the inexplicable feeling that the whole continent has been traversed. Distances are more deceptive than in any country in the world. In the U.S., the vast expanse of the far west can cause mountains that are 30 miles away to appear to be only a short hike. The vastness and spaciousness of the area shorten the perspective. In Ireland the reverse is true. Distances appear greater, but actually are shorter, and the miniaturization of the countryside heightens misperception of special arrangement so that the ever-changing vistas seems to expand in time and space. The little cottages, low stone ditches, small shops, and narrow winding roads fit snugly into green valleys nestling like pebbles in a furrow be-

The village of Belleek from the south. The large building on the right is Belleek Pottery. Cleary's Hotel is the third building from the left with the black trim. The main street of Belleek turns north at the hotel corner. St. Patricks Church, and the top of Breesy Hill can be seen in the distance.

tween great stretches of rolling moorland and turf bog.

The valley of the Lough Erne is no exception, and the twenty-three miles of lake road soon enveloped us with their strange beauty. We drove along lakeside and glens backed by jagged cliffs and bordered by fields of green for which no verbal description is adequate.

The names of places I read off the signposts did justice to the ancient Irish legends and fairy tales—Boho, Derrygonnelly, Carrick, Legg School, Rosscor, Toora, and finally Belleek.

Belleek is situated on the River Erne at the most westward point of Northern Ireland. It is a small village of three or four hundred families, with one wide central street. One approaches the village from the Enniskillen road across a narrow strip of the Free State that juts like a thorn into ruptured northern border, where it lies parallel to the river. Strategically placed on this point of the Free State territory are a pub and a small shop. These are advantageously located for their owners, as the citizens of the North were more prone during the war to buy their cigarettes and spirits at a lesser price in the neutral Free State than in the North with its higher prices. Furthermore, fresh eggs, cheese, bacon, and other such delicacies could be obtained without ration coupons.

It was breach of neutrality to drive an Army jeep across this pointed strip into Belleek. In order to avoid the neutral strip, we crossed the river three miles east of Belleek at Rosscor Bridge, which marks the beginning of the river where it drains Lough Erne.

As we crossed the bridge and curved north we came upon a gatehouse at the foot of a dirt road; above on the hill we could see the antenna poles of the radio range station. The entrance was on the opposite side of the hill along a lane overarched with ancient beech trees. Unlike the coal-blackened beeches I had seen around Stoke-on-Trent, these trees rose like whitened pillars through the early morning sunlight. The tangy smell of peat from a nearby farmhouse filled the morning air. Brown and white cattle drifted in the wake of their leader, browsing along the shoulder of the roadway.

The barracks was long with a concrete floor and two small, round, tub-like coal stoves at either end. It appeared almost empty, for there were only six bunks in it along opposite walls. Large French windows that opened out looked across a flat, grassy cow pasture. From the line of windows on the opposite side was a sweeping view of the Irish moorlands across the border. The boundary of the border to the north of the barracks was marked by the small Keenaghan Lough. Nestled at one end of the lake was a whitewashed Irish farm cottage known as Barnes' Cottage, and at the opposite end, where a small creek drained the lake, was the McGee cottage. It was here that my future wife grew to womanhood.

The Sergeant introduced us around. "This is Rocky Costellano," he said. "And this, Rocky Grassano." These two Americans, both from New Jersey, both called Rocky, became

The "barracks," center right, from high atop an antenna pole. The land is called Magheramenagh for the castle grounds where it is located. Magheramenagh means "high plain" in Gaelic. Far to the west is Ben Bulbin in southern Ireland. This photo was taken by the author while working on the antenna during World War II.

Card advertising Cleary's Hotel

known about the countryside as Big Rocky and Wee Rocky. They were radio range operators, so they pulled shifts at the range station with me the only technician at the site. My main duty as a trained technician was to see that the station stayed on the air twenty-four hours a day, twelve months a year. The fifth G.I. besides me, the sergeant, and the two Rockys, was an extroverted truck driver named Jonas. We had no cooking facilities at the barracks, so twice a day we would go into Belleek for lunch and dinner at Cleary's Hotel.

Cleary's Hotel is located at the end of the broad main street of Belleek, which, approached from the north, appears to dead-end at a large, imposing cream-colored building with several wings and steeply sloping gables. This was Cleary's competitor, Elliot's Hotel (since blown up by the I.R.A.). Cleary's, now only a pub, is on the east side at the end of the street where the road takes a sharp turn around the corner and passes on toward the pottery before turning right again and crossing the ancient stone bridge that in those days humped across the swiftly flowing River Erne. The bridge was exceedingly narrow and impractical as its sides were constructed of high stone balustrades. There was room for only one car to pass at a time. As with many things impractical, the bridge was quite charming and contributed greatly to the scenic beauty of this exquisite town. Alas, as with most things, both charming and impractical, it was also later destroyed and replaced by a concrete monstrosity. Back then, however, one could speculate as to its ancient origin, as the stone balustrades along the sides were constructed so that there were three triangular niches off each side. This type of construction is found in only the most ancient of bridges and the side niches served the pedestrian as safety step-asides during the inevitable emergency as when the helpless foot traveler was caught in front of the galloping horses of the stagecoaches or the local militia.

Advertising for Cleary's Hotel stated that it was both tourist and commercial, and was "most central for fishing, shooting, and visiting all the local places of interest." It further stated that "visitors who come once will call again and recommend it to their friends." With this latter statement, I was inclined to agree, if not for the comfort of the inn, which varied with the coldness of the day, then most certainly for gentle and kindly people who owned and managed the establishment. If one were to list all the virtues a traveler or visitor to a strange land might wish its inhabitants to possess, certainly leading such a list would be kindness.

The stranger deprived of warm and hospitable treatment by unthinking locals soon becomes both suspicious and cynical. The Irish are renowned both for their hospitality, and their ability to converse on almost any subject produces very few cynics among their visitors. Because of the harshness of the Irish rural life, one might expect harsh treatment, but the opposite is the usual in Ireland. The Irish are often moody, gay at one moment, melancholy at another, but these moods are softened with a lively humor, and love of their fellow man.

The Clearys, no exception to the traditional Irish character, were more than patient with the five yanks who invaded their small hotel. The family consisted of Des Cleary, the proprietor,

The Cleary family: Des and Mary with Jimmy and baby Mary

Mary, his wife, and his partially crippled sister, Amenda. Their son Jimmy, three years old, followed the yanks about, forever overturning their cups of tea, and otherwise acting precisely as a three year old is expected to behave. Finally, there was Bridget. Although to all outward appearances a maid, she had spent her entire adult life with the Clearys and was, more precisely, a member of the family. Jimmy was the apple of her eye and, consequently, could do no wrong.

Bridget represented the last of a disappearing type of European woman, for which there is no comparable replacement in our modern world. She was faithful, hardworking, and born to the land as if her body has been molded and cut from the rocky Irish hillsides. In truth, her apparel, often smeared with the black soot of the turf fire, or her brogans, covered with manure and bits of heather from the pasture where she tended the cows, attested to her closeness to the land. During the late summer haying season, she could rake, stack, and thatch the ricks of hay to the shame of any man. Some of my friends thought her dirty, but city bred, they mistook the war paint of man's eternal agricultural battle for uncleanliness. In fact, there are two kinds of dirt: dirt of the soil, without which man cannot survive; and dirt of character, in which the selfish and hard of heart wallow about. Bridget was, in fact, not unwholesome, but to the contrary, saintly in her unselfish devotion to others. Because she was simple of heart, and spoke with the ancient brogue of the Donegal hill country, some also considered her somewhat less than intelligent. Again, nothing could be further from the truth; she knew the techniques of her hostelry and agricultural trades and seldom wasted time on the useless or unnecessary.

We ate on the second floor of the hotel at a huge oaken table in a room that served as both a sitting room and dining room for the establishment. Against the outer wall, between two windows overlooking the bridge and river, was the fireplace. The chairs with their backs to the turf fire were by any measure the ideal spot. The cold chill of an Irish winter can only be overcome by toasting one's backside at the turf fire. In the early days, at the turn of the century, and prior to the war, Cleary's was filled with Scottish and English fishermen on their fortnightly holidays. The River Erne was renowned for both trout and salmon fishing. During the war, however, when times were hard only the odd commercial traveler stopped at the hotel. Invariably, at dinner time, they sought to seat themselves by the fire and just as invariably, Bridget would shuffle through the door with the tea and intervene to reserve the favored chairs for the yanks.

"Shure oi'll be after seein ye sit in this place, shure oi've fergot entoirely to tell ye, this one's Mr. O'Callaghan's place at ta toime."

In Ireland teatime comes often, and for some reason no between meal repast in the world can equal Irish bread and tea at "ta [as in *bay*] toime."

Fair day in Belleek, the 17th of each month.

Belleek On the River Erne   11

At the end of the day we usually ended up in Cleary's kitchen around the turf fire. Amenda, because of her crippled condition, held the place of honor closest to the fire and from her seat directed a communication system, equal to that of the Air Corps, covering the happenings of the entire area from Belleek to Ballyshannon across the border in the south. Although she never moved from that chair, she could at any moment tell you exactly who was courting whom at any crossroad for miles around. Besides keeping tabs on the behavior and idiosyncracies of her countrymen and women, she was also the matchmaker for the yanks.

"Shure I have a nice wee lass for ya, Mr. O'Callaghan. She's after askin' who's the tall yank all the toime."

Amenda also gave details of the history of Belleek and better days when her father was the manager of Belleek pottery. The fame of Belleek china is worldwide; it is considered by connoisseurs of good china to be superior to any of the more delicate varieties of chinaware. It is produced from a special clay and hand made at the potter's wheel. Egg shell thin, it is actually translucent, but it is as tough as china twice its thickness. Because of the fame of the pottery, Belleek has a reputation and renown that belies its small size.

Life in Belleek itself was quiet though not without some excitement. The highlight of each month was fair day, on the seventeenth. On that day the farmers from miles around Belleek would drive their cattle into town to sell. The wide main street of Belleek would become a quagmire of cow manure and the air odoriferous with the scents of sweat and Irish whiskey. Bartering for beef seemed incomprehensible to the outsider, as the shouting of Irish farmers facing one another resembled a violent argument, which, in fact, it was not. Time after time, one man would spit in his hand and slam his hand into the hand of his counterpart with a vigor that would shake them both to the tips of their Wellington boots. When the final thunderclap of hand slapping had died away and the deal was set, the two would adjourn to the pub and seal the agreement with a nip from the bottle.

On fair day the R.U.C checked cattle for clipped ears, as cattle brought across the border for sale were thus marked for custom fees. The price was higher in the north than the south due to the war. A sly Irishman who could sneak a herd across the border at some desolate moorland crossing could make a killing.

The R.U.C often stopped by our barracks for a wee bit of yank coffee while absorbing heat around our pot-bellied stove. One day Sergeant Ryan knocked on the door just as I was getting ready to bicycle into Belleek on fair day. It was a bright sunny day for picture taking.

He pointed out our barracks window. "Are those the same cows that were there last week?" he inquired.

"Same cows?" I asked. "What do you mean?"

"Are they after being the same as last week?" he repeated. "Shure that black and white one wasn't in the pasture last week."

I looked out the window again. "How do you expect me to tell one cow from another? They're like Japs," I said "You know—in the war movies. They all look alike."

"Ah, shure now, cows aren't all lookin' alike. Shure those aren't the same ones that were there last week. Shure ya are no help atal."

As the months passed, I began to study cow configurations and coloring and sure enough, it did seem that the herd changed makeup with considerable regularity. The police sergeant was no doubt correct that the grounds were used now and again for a passing smuggled herd.

The border was a nuisance and of the greatest irritation to the Irish, some of whom had to bicycle seven miles into Ballyshannon to buy their supplies although they lived only a hundred yards from Belleek. On dark winter nights food and goods flowed both ways across the border and little blame could be placed on the locals for smuggling. It seemed that the whole tariff system was set up to inconvenience the common people without serving at all the protective function for which it was originally conceived. The Irish border represented one of the worst examples of frontier bureaucracy in existence. One could be searched three times on a train going from the town of Bundoran in west Ireland to Belfast in the east.

The big event on fair day was the dance in Johnny McCabe's hall, which started promptly at 9:00 p.m. and usually featured a local ceili band consisting of a set of drums and three fiddlers. The ceili dances are the traditional dances of Ireland and in many respects resemble our square dances without the calling. Square dancing is in fact derived from the Gaelic ceili dances. Some of the dances are quite involved and have such picturesque names as "The Siege of Troy," the "Walls of Venice," and the "Walls of Limerick." They represent in dance and music Ireland's old and continuous battle for independence from England. Interwoven with the ceili dances were such physically demanding ones as the Scottish Highland Fling and the slower modern fox trots of the 1940s. The fiddlers were good and considerable energy could be expended during a very few dances. Since the hall was small, and none too well ventilated, bodies were somewhat closely packed and the haze from cigarette smoke rather dense.

The system for meeting lassies and selecting a dancing partner was simple, yet sophisticated beyond any modern folkways of matchmaking. Since the hall was small, it contained only a linear arrangement of benches, whose length of these benches seemed able to accommodate exactly half the occupants of the hall. After each dance and before calling another, the benches were taken up mainly by the men of the crowd, thus leaving little room for the colleens. The arrangement might seem very ungentlemanly to the stranger, but was not intended to be so, for it left the colleens a considerable variety of laps to sit on. The system had several additional advantages as a social custom first, since all the girls were sitting on laps, no one girl could be considered forward and second, it is extremely difficult not to strike up a conversation with a lady sitting on your lap. It was in fact extremely easy to find out exactly who was who about the countryside, for the Irish have few inhibitions when it comes to conversation.

There was, however, one disadvantage to the Irish system of matchmaking, which was that it discriminated against the newcomer, especially one from across the sea and in yankee uniform. The Irish country people, in spite of their great love of America and Americans, were no different from people of any other country in respect to the soldier. A soldier is, after all, a soldier, and well understood to be less inhibited than the average citizen who is not subjected to the unknown terrors which are a part of a soldier's lot. People the world over simply do not want their daughters associating with strange men in uniform. Should a yank occupy a bench, it took considerable courage for an Irish lass to perch herself on his lap. As a result, most of us were often seen leaning against the wall of the hall, not wishing to appear ungentlemanly (by our standards), and not entering into the merriment at all.

It was several months before I worked out a system of my own that enabled me to join the Irish system. I bought a ticket to everything, or for anything being raffled, from the lovely colleens who peddled them around the benches. By never missing a dance (I could trade my night shift to Wee Rocky who was married), I soon got in enough words to become acquainted with a few of the colleens. From then on, they lost their aversion to the yank and things progressed normally. The dancing was fun, and the music haunting, and that the system worked well is attested to by the fact that one of the Irish girls, named Winnie, who came from the McGee farm out by Keenaghan Lough, became my bride a few years later in New York.

One of the Irish lads I became acquainted with at McCabe's Hall was Marty Keegan. He was a tall, gaunt looking Irishman of my age and we were soon close friends. His plans were to join a seminary and eventually become a priest. Because he was plagued with asthma, however, he had not been able to pass the rigorous physical requirements for life in an Irish seminary. He was the son of the schoolmaster at Creevy, and we would often bicycle with the Master to his school, and then continue on to the rocky coast along Kildoney Point where Marty had fisher-

Railroad station at Omagh.

men friends who battled the winds along this rugged coast, setting their nets from small boats for the green speckled herrings. The fish were peddled at stands on fair days in the surrounding Donegal towns. Sitting on the rocks after a day at sea, Marty and I would plan our future after the war. Across Donegal Bay the rugged peak of Slieve League rose in the distance. There, St. Columkille roamed the Donegal highlands converting the ancient Celtic peoples to Christianity. Marty often spoke of St. Columkille and how he was responsible for the spreading of Christianity across Scotland and England, from Iona, the great Scottish center of religion in Western Europe.

We hoped that the days spent rambling over this wild and restless land the exposure to the sea breezes would help Marty's asthma. I went on as many of these bicycle trips as possible with Marty and immersed myself in photographing the intricate rock structure of the Donegal coast. It is the rocks that contributed so much to the scenic charm of Donegal. The cottages are of rock, and the ditches of rock, and everywhere thrusting up from between the green and brown fields or hanging suspended on the edge of sea-sprayed cliffs are rocks. There are all varieties of rocks, from the common limestone from which the farmers get lime for whitewashing their cottage walls, to complex mixtures of igneous rocks thrown into long folds and ridges stretching in waves across the land.

During fall the weather becomes rainy and cold and the sun drops from the sky faster each day that winter approaches. At the little railroad station at Omagh I waited one day in the early dawn for the train to Belfast. The red streaks of dawn to the east outlined in silhouette the flag-like railroad signals at the end of the long concrete platform. They resembled banners flying stiffly at attention in the cold morning air. Two sets of rails, brightly reflecting the morning light, led off into the distance. They were soon vibrating to the squealing wheels of a blackened engine as it filled the silence of the beginning day with the remote hisses of its puffing body. Grey smoke poured forth from the stack, and the smoke, the flutterless signals, and even the drifting clouds all seemed to point to and to direct one's gaze to the west toward the Sperin mountains that lay purple in the early light. Across the tracks was the signal and the gatehouse. Like all Irish and English small-town buildings, it was square at the end, then steeply slanting at the roof, then square again where the chimney sat atop the slope, the pots of the chimney sticking like three pegs into the sky above.

I shivered on the platform for a while, then crossed to the station on the overhead walkway to inquire if the Omagh-Belfast train was late. On the cross-walk above the track, I stopped to watch the little engine, then when the signals, smoke, clouds, and engine seemed in proper harmony, I snapped a picture. I closed the lens down to black out detail of the engine and station but at the same time let in the light beneath the low-hanging clouds. The silhouette I captured would one day bring back the same scene to memory. This is the magic of photography, that in one instant a flicker of life can be captured and held. Never again in the whole history of the world would this precise set of circumstances be brought together. The circumstances would dissolve with time but the record of them was stored in my little black box. I thought what fun it would be to travel the world with the box, not in the army with strings attached, but free to head here or there wherever the will directed. I resolved that when I got out of the army I would tramp the world before I returned to college and a more restricted life. I didn't know how I would accomplish it, but I knew that I would go around the world at least once with my camera before I settled down to study and a profession. There was the Taj Mahal to see, and the Parthenon, and China, and LePuy in France, Rome, and so many others. As a boy I had been an avid reader of that wanderer from Memphis, Richard Haliburton. His book, *The Seven League Boots*, and his *Letters* home to his family, have probably influenced a whole generation to wanderlust.

Presently, my train stopped at the platform. It was an hour behind schedule and my main concern was to get to Nutts Corner in time for a flight that day to London. My furlough had already begun and my plan was to try and locate and visit my brother in England. It might prove an impossibility for although I knew he was with the 507th Airborne Parachute Regiment, where the Airborne were training was top secret. Like me, he was to survive the war, although he was wounded in the Battle of the Bulge. His war was to be far different than mine.

I found my brother despite all security obstacles and I trained for a week with his platoon. I was not to leave Ireland again until about a year after the war in Europe ended. My military occupation had been classified essential. Living in Ireland was much more primitive than life in the United States. I lived well on cabbage and bacon, tea, meat, eggs, and bread; but I could obtain none of the foods in wartime Ireland that we were used to in the States. There were no plumbing facilities in the barracks and to keep warm I had to constantly attend the barracks fire. One might say my standard of living was low and yet, by my own estimate, I had settled in an enchanted place. I was close to the land and the people who worked the land. Surrounding me were miles and miles of wild moorland, mountains, bogs and isolated loughs nestled like sparkling diamonds in a vast mantle of purple heather. It was possible to walk all day and meet only a single person, or to see only one house. At evening, returning home from a hike, the ever-present drumming of the moorland snipe reminded me that I had been transported by a marvelous time machine to a rucksack naturalist's paradise.

2

BUTTERFLY ROCK

I HAVE GOOD REASON to consider Ireland a naturalist's paradise. Like most young people, as a child I was completely in tune with nature. I am furthermore convinced that up to the age of twelve or thirteen, what sociologists call the age of puberty, every living human being is born in tune with nature and is in fact a bona fide mystic. What adults disparagingly call daydreaming in young people is in fact a heightened state of mysticism, a state from which even the Army Air Corps never succeeded in redeeming me.

The first definition of *mystic* in Webster's Unabridged Dictionary is "of or relating to ancient mysteries" (as the Greek Elusinian, or Egyptian). It is Webster's fourth definition that I believe is the best: "inducing a feeling of awe, wonder or similar properties."

Anyone who has any clear certainty of childhood memories, as most people do, recalls hours, usually in the spring and invariably occurring between sunrise and ten o'clock a.m., when the ecstasy of just being alive flows across the body as a soft morning breeze flows between the branches of sunlit leaves. In fact, a gentle spring breeze actually heightens such a happy, mystical feeling. Mark Twain in *Tom Sawyer* and L. Frank Baum in the Oz books wrote of it best.

As the hardships of life increase with age, and indeed rich and poor alike are burdened with hardships, we lose our ability to daydream. However, do not condemn the pragmatist, the politician, the businessman for losing this childhood characteristic, for to dream too deeply is to fail in life and perhaps even to starve one's own family both bodily and spiritually. The starving in the Third World are definitely neither

Scarecrow—the thinker in the OZ books.

daydreamers nor survivors.

A very few lucky people learn to survive and yet daydream for the rest of their lives. They become monks in monasteries, artists, scientists (not scientific technocrats), even writers of children's stories, like L. Frank Baum. Baum was in fact one of the few men who utilized his dreams, his childhood mysticism, to survive.

L. Frank Baum failed at everything he tried, such as running the family petroleum firm, which collapsed miserably after his father died and he took over. When he was twelve years old his family sent him to Peekskill Military Academy to "shake him out of his dream world." That attempt to clear out his mysticism obviously failed totally.

The Baums were always at the edge of financial disaster. Frank would entertain their four small children with homespun tales of adventure. During a storytelling session one of his children asked him the name of the land where these marvelous adventures took place. Frank spotted his file cabinet in the corner. One of the drawers bore the letters O to Z. In a single mystic intuitive flash he answered, the Land of Oz. Thus was born the most magic land ever devised in the mind of an avowed daydreamer. In our petroleum world, Ireland is indeed one of the few remaining mystic spots left on earth. There are, of course, mystic places in every country, such as Hueco Tanks in Texas, Mesa Verde in Colorado, LePuy and Lourdes in France, Luxor in Egypt, Izu Peninsula in Japan, but in the realm of mysticism all of Ireland is magic. That is precisely why the border between north and south is so artificial. There will never be peace until the hard-headed northern Protestant mystics compromise with the equally hard-headed southern Catholic mystics. They work magic against instead of for each other!

When I was a child my mother and father used to take my brother and me picknicking in the huge pile of volcanic rocks called Hueco Tanks (described in my book *Ancient Mysteries, Modern Visions*). Hueco Tanks rises up out of the great southwest desert thirty miles east of El Paso, Texas. It was a sacred

The haunted ruins of Magheramenagh Castle where the radio station was located. The radio beam station was behind the castle to the right. RAF Coastal Command used the beam to home on after sinking German subs in the wild north Atlantic.

spot to the southwest Apache and Navajo Indians. When I was eighteen years old, just prior to joining the Air Corps, I hiked out to that curious pile of rocks and spent the day daydreaming.

My hike to Hueco Tanks was the last time I ever saw that magic spot until forty years later when I returned again with my wife. From that pile of rocks in the desert to the pile of rock in Ireland was for me but one short trip. A day at Hueco Tanks was spiritually very little different from my years in Ireland, for one of the characteristics of the mystic experience is that there is no real sense of time. The relativity of Einstein's time is not proved by mathematical formula but rather by the timelessness of dreams, and the mysteriousness of life in the sun.

Scattered about among the green humpbacked dunline of the Erne valley, where I lived during the war, are numerous ancient castles, some crumbling, some still occupied. Castles are places for dreaming, especially dreams of knights in armor, of fair maids, and the sport of falconry. As a teenager I was one of America's first practicing falconers.

My rambles around the countryside of Ireland led me to the conclusion that there is a subtle magic force in the stones of old castles similar to what I felt as a child in the stones at Hueco Tanks. Why is that desert rock pile near El Paso an oasis of green plant life? The rock "tanks" that trap the scarce desert rain must be a part of the answer, but that does not explain the plant vitality of the spot, nor the feeling of oneness with life that it imparts to all but the most most insensitive visitor, nor does it explain why trees grow from the ancient castle walls.

I believe in inherited memory, a slightly different concept than reincarnation, but nevertheless a sort of mind reincarnation, along genetic lines. It is fashionable nowadays to call such a phenomenon sociobiology, the inheritance of cultural values. If we label it sociobiology it is scientifically acceptable; if we call it inherited memory it smacks of mysticism. There will always be a certain amount of hypocrisy in science as there is in every human endeavor.

Just as I have loved birds of prey from my first childhood vision of bald eagles in the big wire cage of the Memphis zoo, so also I seemed to have inherited a genetically linked attraction for castles and cliffs. How else can one explain the magnetic attraction to falconry, castles and climbing by a boy born in the coastal plains of Georgia in 1923?

The medievalism of my being was fixed in my inherited memory long before I ever began to read on such subjects in the public libraries of Denver, Detroit, Albany, New York, and Texas where I spent my formative high school years.

In all the world the gods of fate could not have chosen an environment more suitable to my character than Fermanagh, County Ireland. It was as if, in the midst of the most horrible war in the history of mankind, the angels of goodness had led me by the hand and given me a short period of time in the last, agriculturally speaking, western medieval land of existence. However, it was my love of falconry that led me to rock climbing, for falcons nest on cliffs.

Rock climbing for the novice can be frightening, but fear only further accentuates the inspiriting feeling that follows a good climb up a rock face. It was not until a day spent on Ireland's Eye at the end of the war that I really began to understand that climbing up a cliff or rock wall infuses one with a special sort of energy. At one time I believed that the "climber's high" was merely the usual wave of well-being that follows a difficult undertaking. On Ireland's Eye I learned that such is not the case. It happened at the round granite Martello tower that rises up on the west end of a unique camel-backed island near Dublin.

The tower castle on Ireland's Eye off Howth Harbor is historically of no importance, and yet its eight-feet-thick granite walls invoke in one intense vision of medieval conflict. This tower, built on the wave-battered rocks at the western tip of Ireland's Eye, guards the entrance of Howth Harbor. It is the only man-made structure of any permanence on the island, other than a tiny abbey with its so-called leper's window. None of the Irish fishermen I have questioned at Howth seemed to

The O'Shea cottage on the Belleek road. The author's wife stands in the doorway of her sister Agnes' cottage—a great place for story telling in World War II.

The Martello tower at Sandycove, south of Dublin, where James Joyce lived for a very short time with the poet Oliver St. John Gorgarty. It is now the James Joyce museum.

know anything about the squat stone tower. It is indeed a "round tower" but no kin of the Irish monk's round tower. It is short and dumpy like a sway-backed mare, not gracefully tapered and slender like a Celtic maiden.

The squat, short towers, called Martello towers, were built around the edge of Dublin Bay by the British during the Napoleonic wars, supposedly as castle defenses against a possible French invasion.

I first visited the tower in 1945. Much later, reading about the self-exiled Irish writer James Joyce, I discovered that Joyce had written a portion of his semi-autobiographical story, *A Portrait of the Artist as a Young Man*, in a similar tower near Sandycove south of Dublin. At the time Joyce roomed with another famed Irish writer, Oliver St. John Gogarty.

Oliver St. John Gogarty leased the tower for £8 a year. The British had wisely decided the towers were of dubious military value and at the end of the 19th century most were occupied by young men of poetic nature attracted by cheap rent and the allure of seaside life. Joyce apparently boarded with Gogarty for a very short interval in September 1904—not long enough to write much of anything. Today, the Sandycove Martello tower houses a most delightful museum.

The tower on Ireland's Eye off the coast of Howth Harbor is another story. It is a deserted chunk of granite that holds for me an epic memory of my delightful Irish adventure during World War II.

As soon as the war ended in May of 1945 I crossed the border in uniform, having resolved to be the first American soldier in uniform south of the border—as indeed I was. With two Irish friends, Jim Shute and Joe Duffy, I spent a week sampling the pleasures of O'Connel Street and exploring the environs of Dublin. Toward the end of the week we ended up in Howth, the lovely little peninsula that projects into Dublin Bay north of the city. I swam the mile from the harbor wall to the north shore of Ireland's Eye on a bet from one of my companions.

The following is an excerpt from my notes of that day so many years ago:

> The sides of the fortress on Ireland's Eye were extremely rough and of broken masonry. Jim gave me a boost to a ledge and I climbed up from handhold to handhold until I reached a window 20 feet above.
>
> "What's in it?" Jim shouted from below.
>
> "It's too dark, I can't see." I shouted back. Just then the wind took my overseas cap and swished it into the darkness of the tower.
>
> "I can't go back without my hat!" I shouted. "I'll have to see if I can climb down inside and get it."
>
> The masonry on the inside was uneven, as that of the outside, but I couldn't see where to put my hands or feet. After a while, my eyes got accustomed to the darkness and I inched my way down to the rocky floor of the tower. I found my hat, but there was little to explore inside the tower—just a large empty room. I wondered what purpose it had originally served. When I looked up I was suddenly overcome by what might be termed "dungeon fever." The psychology of looking up from the bottom of a deep pit to a little streak of light in the window far above was, to put it mildly, unnerving. I could envision the middle ages when prisoners were lowered into such holes and left to die slowly over the years, being thrown only a piece of bread now and again. Visions of Victor Hugo and his fearful descrip-

tions of imprisonment raced through my head. The window, which from the outside appeared to be only 20 or so feet above the rocks, from the darkness inside looked 200 feet above. I suddenly had a sinking feeling in my stomach. What if I wouldn't be able to find the same handholds going up and was trapped in this musty, rocky dungeon?

I could hear the muffled roar of waves dashing against the rocks outside, then a ghostly stillness followed by another against the wall of this rocky prison. The broken rocks on the floor began to take on shapes and soon I imagined myself surrounded by skeletons. I had not realized coming down how damp and slick the rocks on the inside were. I had lost control of my emotions and the climb up became more and more difficult. I grasped from side to side for holds, but each one seemed to lead too far to the left or right of the small window above. No good to climb to the side of the opening. The higher I went, the further away the light appeared. I could see that the sun was dropping fast and I began to panic—what if night overtook me inside? I had no desire to spent a night in this dismal hole with the squeaking bats above and encircled by water-dripping walls. If I slipped and fell to the rocks below I would probably be killed.

Climbing a rock wall in the dark is nothing like climbing a cliff in sunlight and even cliff-climbing, without ropes, is foolhardy. The young often do things that are dangerous and foolish, but up until now I have always considered myself rather sensible in such matters. I wouldn't climb to a falcon nest in the middle of the night, so why was I hanging on the side of a slippery stone wall in almost total darkness? After what seemed an hour, I grasped the window ledge and pulled myself up. I sat for a few minutes in the window, then climbed easily down the outside wall. Jim Shute could see I was tired.

"Was it far down inside?" he asked.

"No, hardly any climb at all," I replied with some bravado. I knew that what I said was only partially true. It probably really wasn't much of a climb, as climbs go, but it was the psychological fear of being trapped in a hole that had been unnerving. For the first time, I knew how Becky and Tom Sawyer felt, lost in the labyrinth in a lightless cave. We rowed the dinghy back to Howth and caught the train at Dublin Station for Belfast and Belleek.

The incident inside the Martello tower is as clear in my memory today as it was in 1945. Why? Mainly, I suppose, because it was a marvelous day at the seaside with two very close Irish friends and the last day of our Dublin vacation, it is imprinted in my mind. However, despite the joy of that day on

O'Connel Street in Dublin. Nelson's pillar erected in 1808. At 1:32 a.m. March 8, 1966 the pillar was blown up by persons unknown. There were no cars in 1944.

Ireland's Eye there was a subtle sort of mysteriousness about my hour inside the Martello tower that I could never explain. I have been a cliff climber since my earliest boyhood. Cliffs and high rock places are to me as a ten penny nail to a magnet, and as a youth I had considered my most pleasant days those spent scrambling up to falcon eyries, something unfortunately denied to today's young falcon and cliff lovers. There are now laws against disturbing falcon eyries.

As a youth, of course, I knew nothing of the techniques of rock climbing and often got myself in the most terrible fixes, of which the incident at Ireland's Eye is by no means the most disconcerting.

At that time there were few books on rock climbing, so I learned most of my techniques by trial and error. Somewhere along the way I did obtain what I consider a Victorian master-piece, *The Art of Mountain Tramping* by Richard W. Hall. It is filled with what Mr. Hall calls British mountain "Walks and Scrambles." It has chapters on everything from "Stockings and Macks" for outdoor wear to "Bad Weather" and how to get across "Farm Walls, Fences and Gates" in the most gentleman-ly manner, and without upsetting the farmer or landowner. It is, however, the descriptions of his rock scrambles that always leave me enchanted. The name places of his favorites, Grassmore, Griesdale Pike, Great Gable, Tarn at Leaves, etc., lend the book a flavor no modern treatise can match.

There are certain books that one simply does not lend to anyone. Richard Hall's *Mountain Tramping*, Gilbert Blains's *Falconry,* and T. Gilbert Pearsons' *Birds of America* are for me three such books. You may borrow or steal my car or TV, but please leave those cherished books on my shelf! We have books on the joy of everything from sex to cooking, but what is real-ly needed now, during these times of electronic word process-ing, is a book on the "Joy of Reading." Some electronic imbe-ciles are even suggesting that books are outmoded and should be replaced by tapes and green glowing video screens—pity the poor future historian.

The beauty and attractiveness of books is no less inspiring

than the beauty and attractiveness of rocks, and, indeed, without books who would know very much about rocks at all! I should hate to carry video equipment into the byways of the world to study my rocks.

A good book can, of course, effectively tranquilize one. It can hold one spellbound and oblivious to all else, so also, strangely enough, can climbing the side of a cliff or mountainside. I call it the "Joy of Clinging" and up to now no mountain climber has ever given a good explanation for the subtle delight.

Looking back, I believe that it was inside Martello tower of Ireland's Eye that I first realized the power in certain types of stone. It is not the tranquilizing power of a body-doping drug, but that which leaves one calm and peaceful, yet simultaneously with a mind and body energy sustainable over long periods of time. There are many drug tranquilizers on the market that will relieve stress and calm one's apprehensions, however, all these drugs leave one's body as limp and flagging as a headless snake.

I vividly remember that short climb up the slippery side of the Martello tower. As soon as I left the floor of that dark hole, and pressed my body against the cool granite of its thick walls, my panic subsided and "dungeon fever" was cured. As if by some unseen magic I suddenly became as calm and energetic as an Olympic runner. I knew on that day that the elation I always felt when I pulled my straining body up on to a ledge where a wild falcon or eagle nested was due less to the beauty and power of those rare creatures than to the *power of the rock* from whence they nurtured their offspring.

What is this power which is in rocks? Climbers know of it but do not understand it. The ancient Hebrews and Egyptians understood and utilized it—the Jews do to this day at the Wailing Wall. I have never been to that tranquilizing wall of stone, but on several occasions, while traveling around Ireland, I have come across magic rocks. One of these rocks seems to possess all the mystery that prompts mystics such as William Butler Yeats, the great Irish poet, to invoke the spiritualist's concept

The tower house (castle) called Thoor Ballylee (near Gort County Galway) that William Butler Yeats restored and lived in for many of his mid-years (1919-29) during his summers (see chapter 2). It is now a museum.

of life and the afterworld.

From times as distant past as Egyptian civilization, and presumably long before, man has attributed certain mystic and supernatural power to stone, and to crystals which are, after all, a special type of stone.

In mid-life, Yeats went to live in a stone castle, Thoor Ballylee, near Gort, County Galway. On the wall of that Norman towerhouse, as such castles are called, he carved in stone:

> I the poet William Yeats
> With old millboards and
> sea-green slates
> and smithy work from
> the Gort forge
> Restored this tower
> for my wife George
> And may these characters
> remain
> When all is ruin once again.

Sea-green slate as I measured later is one of the most magic of all rocks.

Although I am intrigued by the mysticism of Yeats and admire the originality of such poets as Thomas Kinsella in breaking away from what Maurice Harmon, the writer-critic, called his "Irishness," neither of these poets reach the taproots of my Celtic ancestry as does Patrick Kavanagh. Perhaps that is because I am an agriculturist, and Kavanagh wrote mainly about the small Irish farms. More than likely it is attributable to the phenomena that I call inherited memory.

Which puts Kinsella at a disadvantage where my poetic preference is concerned, for the "Irishism" which Kinsella rejects is a part of my ancestral memory, and who among us wishes to suppress our brain-stored nostalgia, especially of Ireland's rock strewn farms, and rock walled cottages?

Lest my readers conclude that inherited memory is an impossibility, consider the American Monarch butterfly, or the Red Admiral of the Irish hedgerow. The Monarch winters in

Slieve League from "One Man's Pass" where Butterfly Rock is located.

Butterfly Rock.

Louisiana and Mexico and migrates north to Canada where it gives birth to its yellow and black striped caterpillars. Two or three generations later its great, or great, great grandchildren fly back in the winter to the same little forest in Louisiana or Mexico. Although not as spectacular, the Red Admiral performs a similar feat in relation to Ireland and the continent. Now, this elegant process requires a lot of stored navigational memory passed on from generation to generation with considerable exactitude. Scientists may call such natural phenomena what they will, but by any other name it is most certainly inherited memory.

A very special rock I stumbled on clings to a mountain in Donegal where I have observed traveling butterflies stopping to rest. This rock is probably programmed in their inherited memory.

The climb to Slieve League is along a steep ridge on a track called One Man's Pass. The hiker looks straight down a thousand feet from either shoulder of the pass, yet I have seen even the timid ignore the danger and pass cautiously along the scary-named trail. It is, without doubt, the rocks that comfort the fearful. The stones and heather of Slieve League paint a picture of God, and who can be afraid of walking with God?

It is said that the color of the stones of Slieve League changes with the weather that blows inland from the restless Atlantic. That is not true, for the rocks, regardless of the weather, are always green or turquoise, rust or reddish-orange. It is the depth of color, the subtle tint, that varies with the light. The greens may shift from turquoise to spring-leafed green, or the reds from crimson to Irish flag-Orange, but never do the rocks of Slieve League turn to white or grey like the cliffs of Yeats' beloved Ben Bulbin across Donegal Bay.

Butterfly rock is perched beside one of the few flat spots along the trail and looks, as do all the rocks of One Man's Pass, as if at any moment it might slide down the steep mountainside and disappear into the roaring Atlantic a thousand feet below. I certainly hope it never does, for I am fond of that rock. There is no doubt at all in my mind that the butterflies of Donegal need that rock as a resting stop just as I need a bed and breakfast when I wander in Ireland or as I needed Cleary's Hotel during the war.

All over the world there are special rock places where migrating insects congregate. The tops of Tors on the Dartmore swarm with small insects in the summertime. The Bogong moth of Australia congregates in the granite outcrops of Mount Gingera, and certain species of lady beetles spend their winters in the boulder strewn slopes of the Rocky Mountains.

Slieve League has the only butterfly rock that I have stumbled on in my Irish travels; however, the pretty wanderers do not congregate on the rock. It is just every once in a while that a weary individual stops for a few minutes to spread its wings in the sun—almost as if it is absorbing some mysterious energy from the warmed stone.

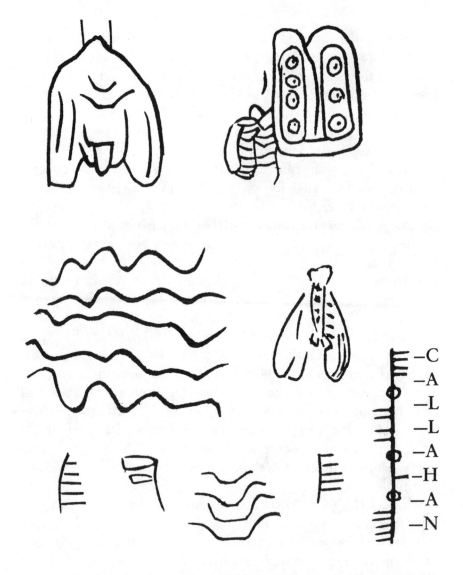

Figure 1. Woodcut of rock carved insects, Oghan Script is author's name (added).

The ancient Celtic and megalithic people knew all about the energy given off by certain types of granite and other igneous or metamorphic stone of volcanic origin. The ancients also knew what the insects knew, for carved into the strange array of megalithic monuments at Carnac on the coast of Brittany in France are accurate carvings of beetle and moth species. The insect figures are surrounded by wavy lines of energy. These stone insect carvings tell me that the megalithic people were as observant in their descriptions of nature as any modern scientist.

The moth on the rock at Carnac is a species in the distinctive family of tortricids. A most unique species is the Green Tortrix, *Tortrix viridana,* which feeds on oak trees in late May. They often defoliate oaks to the extent that the droppings of caterpillars falling on the ground make a pitter-patter sound throughout the forest. The adult Tortricidae at rest have unusual tortoise shaped (rounded) shoulders and sit with their front pair of legs straight out in front of their heads. The antennae are folded back under their wings.

The beetle on the megalithic stone from Carnac is the common May beetle. The esteem in which the ancients held oak trees and the month of May needs no explanation here.

The *Encyclopedia Britannica* says that the May beetle on the slab from the alley stones of Locmariaquer, is an octopus. The elegant sculptured moth from a megalithic mound slab is not identified in the encyclopedia. Since most archaeologists are not naturalists they have misidentified these stone messages.

That the Celts understood rock energy there is little doubt. Ireland is strewn with place names containing the Irish word for rock, carrick. There is Carrickfergus, Fergus's Rock in County Antrim near Belfast, and Carrickoris in County Offaly. Carrickoris, Fheoris's (pronounced oris) rock, sits atop Carrick Hill about three miles from Edenderry, and close by my favorite Irish hiking trail—the tow path of the Grand Canal. One can fill pages with the carrick places of Ireland.

The early Celts of Ireland also understood that the granite stones of Ireland emitted a healing force. Granite is highly paramagnetic (attracted to a strong magnet) and all over Ireland are standing stones utilized for healing and also as birthing stones for the Celtic lassies.

In my war days in Donegal and Fermanagh I had a special stone on Breesy Hill near Belleek where I went to lie in the summer sun and revive my spirits. Breesy Hill rises up in the middle of the Pullen, the rugged moorland northeast of Bal-lyshannon in Donegal. An hour or so on Breesy Hill would put me in touch with God better than an hour in any church—and why not? Breesy Hill was built by God and churches are built by man.

There are more magic stone structures in Ireland than anywhere else in the world. Besides the healing stones there are fairy rings and stone forts, and of course the strong walls of deserted cottors cottages, inside which wild and well-hidden gardens grow in profusion, stimulated by the surrounding rock energy.

The landform of Ireland was shaped by volcanic activity. It is just such violent birth that begets form. Nevertheless, in na-ture, more often than not, form and beauty emerge from violence as one and the same. Like armless Venus, the rock statue of love, the bare shape of Ireland's countryside was revealed when the covering of her ancient oak forests was stripped away by the early Celtic agriculturists. What emerged was a seductive countryside with an appealing form at least as alluring as the Greek goddess. Be that as it may, it is not alone the stark uncovered sweep of Ireland's rocky countryside that seduces one. It is just as mysteriously the magic power of places like butterfly rock—the butterflies know that and so do I. I discovered it many years ago inside the ancient Martello tower on Ireland's Eye.

3

THE ANCIENT LAND OF MIODHBHOLG

THE FOLK MUSIC of Ireland is sentimental and effervescent. The fiddle, tin whistle, boron and ceili songs of old Ireland soothe the ear while simultaneously and mysteriously vibrating the body with waves of nostalgia. A competent artist with an instrument as uniquely simple as the tin whistle can transport one body and soul to the countryside of Ireland. I call the strange moods excited by Irish melodies "musical telekinesis."

Even the winds from Donegal Bay that moan and dance about the upright boulders of granite and schist that top the hill called Breesy, north of the pottery town of Belleek, bespeak the enchanted tones of Ireland. Breesy Hill is a huge rock-strewn magic flute and the flutist is the west wind that blows from distant Donegal Bay.

My many hours spent on the summit of Breesy Hill have led me to believe that there is much more involved in the magic of Ireland than the winds and sounds produced by melancholy

Breezy Hill and the Pullan (see map) in Fermanagh County, from Barnes' cottage on Keenaghan Lake. The view from my barracks window.

weather and her people. Like butterfly rock and the Martello tower the magic lies in her form and her stone.

Breesy Hill is located in the Parish of Carn in the center of a triangular shaped area of rocky hills and bog land called the Pullan. The Pullan, which means puddles, is bordered on the north by the Donegal town of Ballintra, on the south by the town of Belleek and Ballyshannon, and on the east by the little town of Letter. A large area between Breesy Hill and Letter is cut from the eastern half by the border of Northern Ireland. This part of Northern Ireland, as pointed out in Chapter 1, lies south of Southern Ireland.

Since 1969, when the modern "troubles" began, there has been very little violence on the Donegal-Fermanagh border with the one exception that the beautiful Carlton Hotel, at the end of the main street in Belleek, has been blown up—an occurrence as mysterious as the Pullans itself, since the hotel was owned by a Catholic family.

In ancient days the most southerly part of the parish of Carn, where Belleek is located, was called the land of Miodhbholg, a Gaelic word meaning Central Hollow.

According to a short history of the parish of Carn by Father P.O. Gallachair, the first Irish speaking peoples, thought to be the Ulaidh, might have approached the green valley of the Erne from the north while hunting or foraging across the Pullan. Even as late as World War II the misty highlands north of the River Erne were a rich area of game. I myself have hunted snipe, ducks, and even the less common red grouse from its lakes and heather covered hills. I was never fond of hunting with a gun, only with a falcon, but snipe and red grouse were a welcome change from wartime rations.

In my wanderings around the Pullans I grew to love the area and especially the beautiful Erne Valley. The Ulaidh themselves must have believed it a magic spot for as Father Gallachair so uniquely puts it, they looked down from the highlands on the "shelving scene below" and named it Miodhbholg.

Viewed from the summit of Breesy Hill the "Central Hollow" of the River Erne is indeed a rare and inspiriting part

Turf being cut on the Phelan. The bricks of turf are sliced at an angle from the turf bank by the long thin slane. The slanted cuts are visible in the bank. In the background is a pile of cut turf stacked for drying.

of a small portion of the landscape of Ireland. June 23 is the Vigil of the birth of St. John the Baptist and on that day the Irish for miles around Belleek, in days past, climbed Breesy Hill. It was an ancient tradition and those who reached the summit were wont to say that the view is so clear that you can "see them plucking the geese in Scotland."

From the summit one can also see at least seven of the one hundred or so lakes that dot the Pullans. To the north lie the Blue Stack mountains of central Donegal, and to the southeast spread the silver waters of Lough Erne and the giant cliffs of the Fermanagh escarpment which edges the long narrow Lough.

Far to the southwest the prow-like cliffs of the flat-topped mountains of Ben Bulbin jut out above Sligo Bay and the Atlantic Ocean. The mountain reminds one of an ocean liner on the upward thrust of a huge wave, with its prow pointed straight at America. In the path of the rocky ocean liner is the village of Drumcliff where William Butler Yeats is buried. After the war his body was exhumed in Italy, where he died, and carried to Drumcliff. It was the last request of the Irish poet that he be buried between Ben Bulbin of his boyhood and the wild shores of the Atlantic.

The ship-like bulk of Ben Bulbin separates Sligo and the beautiful valleys of Glenade and Glencar from the valley of the Erne and Lough Melvin. Lough Melvin is connected to the ocean by the River Drowes which flows west parallel to the River Erne.

If there is one physical characteristic that dominates this unique area where the Erne and the Drowes rivers meet the sea, it is the simple fact that it demonstrates better than any other geological region how the rocky skeleton of Ireland was formed.

Tens of thousands of years before the Ulaidh people approached the land of Miodhbholg from the highlands to the north, the area lay under the crushing weight of a massive glacier. Glaciers are ice-age time machines, time being measured by geologists in millions of years. The three latest geological

Belleek from the north. The main street is far right, and the back of Belleek Pottery to the left. Far to the south is the battle ship prow of Ben Bulbin. William Butler Yeats is buried at the base between Ben Bulbin and the sea.

TABLE 1
GEOLOGIC TIME CALENDAR

ERA	PERIOD	EPOCH	YEARS AGO—APPROX. START OF PERIOD	ANIMAL LIFE
CENOZOIC				
	QUATERNARY	RECENT	10,000 years	Civilization begins
		PLEISTOCENE	2 million	Rise of human species
	TERTIARY	PLIOCENE	12 million	Mammals, birds, and
		MIOCENE	25 million	insects dominate the
		OLIGOCENE	36 million	land
		EOCENE	58 million	
		PALEOCENE	65 million	

ROCKY MOUNTAINS FORMED IN CRETACEOUS PERIOD

ERA	PERIOD	EPOCH	YEARS AGO—APPROX. START OF PERIOD	ANIMAL LIFE
MESOZOIC				
	CRETACEOUS		135 million	End of dinosaurs; *second great radiation of insects*
	JURASSIC		181 million	First mammals and birds
	TRIASSIC		230 million	First dinosaurs

APPALACHIAN MOUNTAINS FORMED IN PERMIAN PERIOD

ERA	PERIOD	EPOCH	YEARS AGO—APPROX. START OF PERIOD	ANIMAL LIFE
PALEOZOIC				
	PERMIAN		280 million	Expansion of reptiles; decline of amphibians
	CARBONIFEROUS		345 million	Age of amphibians; first reptiles; *first great radiation of insects*
	DEVONIAN		405 million	Age of fishes; first amphibians and *insects*
	SILURIAN		425 million	Invasion of land by *arthropods*
	ORDOVICIAN		500 million	First vetebrates
	CAMBRIAN		600 million	Age of marine invertebrates

The Ancient Land of Miodhbolg 47

eras, the Cenozoic, Mesozoic and Paleozoic ages, stretch back over 600 million years. Fossils began to appear in geologic strata during the Cambrian period of the Paleozoic age. The Cambrian is known as the age of marine invertebrates as much of the present land mass, including the Pullans of Ireland, lay under water. There are very few Pre-Cambrian fossils, so it is almost impossible for geologists to date rock before the Cambrian era. We might say that the Cambrian is the beginning of "rock time" as geologists know it today (Table 1, page 47).

During the Paleozoic age Ireland lay under a vast sea that stretched south westward from Scandinavia across the northern portion of England. Very little is known about the older pre-Cambrian rocks that lay beneath the sea. Upon the sagging rock base of the sea, sand and clay from nearby land masses washed, first filling and then insulating the rocky underbase from the cool waters of the ocean. Simultaneously with this long slow insulating process the land masses fractured and drifted apart.

The theory that describes the movement of the continents was the brainstorm of German meteorologist Alfred Wegener. In his book *The Origins of Continents and Oceans*, Wegener suggests that continents are not stationary but drift apart over the eons on an undersurface of semi-solid igneous rock called the *sima*. This was the first description of what today is called "continental drift."

The book was published in 1915 and some biologists were captivated by Wegener's theory; most geologists, however, ridiculed the idea. It is, after all, difficult to imagine the huge land masses floating about and gradually drifting apart like so many logs floating in a bog.

New scientific theories are always based on ideas by imaginative and observant individuals. For that reason ridicule and controversy are the usual response. Science would not be science without criticism. The negative extreme of criticism is ridicule. Criticism is the fuel of emotional energy that drives the imaginative scientist onward to prove his bold ideas. Criticism is analagous to the pressure of sand and clay that pressed down on the rock of the underlying sima until the compressed stone heated to explosive force. Thus are underwater volcanoes formed and thus also do new ideas explode into the collective consciousness of mankind.

We may understand, then, that it was not a geologist who discovered continental drift but a meteorologist. He based his thesis on the fact that fossil plants endemic to tropical regions were found imprinted in rock as far north as Alaska. His conclusion was that the continents drifted like huge rafts from the more southern regions, where the plants grew, to the far northern waters.

Geologists and paleontologists calculate that the sea that covered Ireland came into existence about 700 million years ago during the Precambrian era and was in existence until about 400 million years ago during the Devonian period of the Paleozoic era. During that vast stretch of time sea creatures with external skeletons of calcium carbonate began to develop. It is the shell remains of these fossilized sea creatures, preserved in the great mass of sediment, that enable paleontologists to date and describe rocks from the Cambrian period.

Millions of years after the Cambrian age of invertebrates, forms of life resembling present day insects began to appear in the rocks of the Devonian period (Table 1, page 47). The Devonian period, which began 405 million years ago, is considered the beginning of the age of fishes, amphibians, and insects. The oldest order of insects still in existence are the Collembola, tiny creatures with the common name of springtails, that occur in grasses and rotting organic matter.

While the forms of these fascinating living creatures were slowly evolving, so also was the geologic form of the great land masses.

As the insulating layer of sediment pressed down on the undersima of molten rock, the earth's heat of radiation, unable to escape into the atmosphere, stored up explosive forces that triggered devastating volcanic eruptions. These natural earth explosions took place mainly along the edge of European land

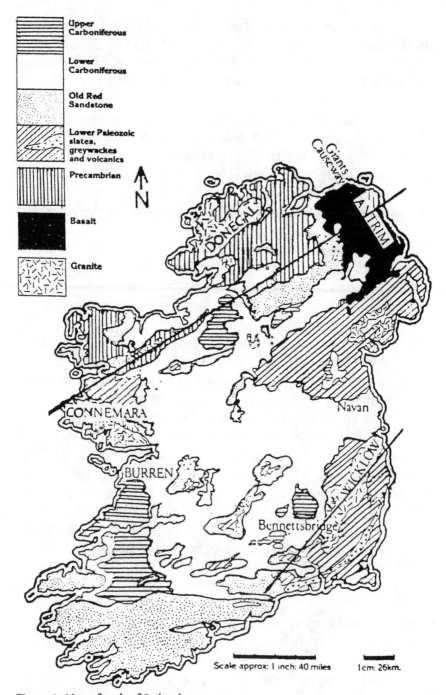

Figure 2. Map of rock of Ireland

formations. In the case of Ireland it resulted in a tremendous volcanic fault that lies along a line from western Galway to the northeastern tip of Northern Ireland (present-day Ulster).

As the sediment from land erosion grew in mass and weight, it compressed fine clay particles into shale and cemented sand grains into what is present-day sandstone. The molten rock called magma, squirted, so to speak, to the surface and solidified into the numerous igneous (fire formed) rock and stones that cover the surface of Ireland.

Geologists refer to this period of land formation as the Caledonian upheaval (see figure 2, page 50.) They envision that the crust of the earth northeast of the volcanic activity line crunched, like the jaws of a vise, against the line to the southeast, compressing and uplifting the sediment that lay between them. During this period there was tremendous upward rock folding and land formed where formerly there was sea. A smaller parallel, but similar fold (see figure 2, page 50) resulted in the Wicklow mountains south of Dublin. It is called the Leinster volcanic axis.

The Caledonian upheaval that produced Northern Ireland also reached across what is now the North Irish Sea and thrust upward the northern land mass of Scotland. Even today in many ways the geologic land from Scotland resembles that of Northern Ireland.

The next fifty million years were an era of time significant to the history of the agriculture of all Ireland. After the Caledonian upheaval, Ireland was left above the surface of the water. The sea, however, still covered southwest England. Since Ireland was no longer under the sea, few fossils have been found from this period; however, the silt deposits of Devonshire England, which was still under water, are rich in marine fossils. The Devonian period lasted for over sixty million years and ended about 345 million years ago at the beginning of the Carboniferous period, considered the age of the first great radiation (movement) of insects.

The Irish highlands to the north and east were eroded by time and alternating extremes of temperatures. Sand and gravel washed from the mountain by rain and rivers spread south in wide belts across the lowlands. Most of the eroded materials were made up of small grains of quartz containing iron particles. Over the ages the iron contained in the mixture oxidized, turned red, and was cemented into the quartz sediment, resulting in the formation of what geologists call old red sandstone.

Slow earth movements ripped across the little island. The highlands continued to rise and the lower basin to sag under the tremendous weight. The sandstone formations grew to tremendous thickness.

On a modern-day geologic map of Ireland formations of old red sandstone are seen as oriented in bands stretching from northeast to southwest. The formations lie in a broad belt between the northern volcanic fault and the Wicklow volcanic fault.

Beds of sand that have been consolidated into rock are usually lumped under the category of sandstone. Other sedimentary (deposited) rocks are shale (consolidated clay), conglomerate (consolidated gravel), limestone (consolidated shell creatures), and dolomite (a special form of limestone).

About 345 million years ago, during the Carboniferous period, portions of Ireland were again under water and the Wicklow highlands were connected with the highlands of Wales, forming a large island called by geologists St. Georges Land. During this period very little sand was washed from the highlands into the underwater portion of Ireland. The water was warm and suitable for the growth of corals and other sea creatures rich in calcium carbonate. The entire central lowlands of Ireland, which lay under the sea between the north and south highlands, was transformed into the calcite limestone (consolidated sea creatures) that today forms the fertile central bowl of Ireland. During the Carboniferous period the rate of sag of the sea floor slowed and calcite rich sediments grew to greater and greater thickness as sea creatures evolved and dried over the eons. The water became shallow and eroded rock (sand) washed from the surrounding highlands, forming deltas that pushed out into the sea. The slow, steady erosion of land

Glendalough round tower looking to the west. The monastery was founded by St. Kevin who died in 618. It is considered the first university in Europe and is by far the most beautiful round tower site in all of Ireland.

and flow of sand into the delta resulted in thick deposits that were slowly transformed by pressure into sandstone.

Toward the end of the Permian period a second major earth ripple called the Armorican upheaval thrust upward and folded the sedimentary area of southwest Ireland against the rocks of the north. It left Ireland high and dry.

As forming deltas enlarged they grew upward, forcing the streams and rivers to cut through the newly formed land. Swamps blossomed and rotting organic substances matted the land, forming dense tropical forests. Huge amphibians began to move from the sea to walk and crawl on land.

The appearance of dinosaurs marked the end of the Permian period and the beginning of what geologists called the Mesozoic era. The formation of the Appalachian mountains in the U.S. is considered to have occurred during the Permian period. The Permian separates the Mesozoic age of dinosaurs from the earlier Paleozoic age (Table 1, page 47).

Today, the landscape of Ireland shows the results of the great Armorican upheaval. The folding is apparent in areas where the carboniferous beds directly overlie Devonian sandstone, e.g., the valleys of Glencar and Glenade. The later carboniferous stone near Sligo is present in the valleys and the earlier Devonian sandstone appears along the higher ridges where it was thrown up.

The great Armorican upheaval was fortuitous not only for that reason but also, as G.A.J. Cole has pointed out, had the Devonian rocks been limestone instead of red sandstone and the later carboniferous limestone, a harder quartz, the fertile lowlands of Ireland would not exist. The geological reasoning behind Professor Cole's thesis is complex but in general terms the parallel ridges of old sandstone, being more resistant to weathering and erosion, allowed the softer shale and sandstone to fill the lower basins and acted, so to speak, as hardened, corrugated funnels for eroded rock that slowly disintegrated into the soil which now overlies the limestone base of Ireland.

We may understand then that the tremendous geological forces taking place over the eons and up to the great Armorican upheaval were setting the stage for Ireland's modern role as a primarily agricultural country.

The trapping and funneling effect of the water-resistant old sandstone ridges, the grinding and fracturing of the softer clay and stone to form soil, and the mixing of that soil in the deltas and lowlands with the rotting organic matter of the Mesozoic age of the dinosaurs was setting the stage for the geological evolution of the Land of Miodhbholg—the valley of the Erne.

It was into this land, where the Erne and Drowes rivers meet the sea, that the first Ulaidh moved. It was also into the same "shelving land below" that I was thrust, thousands of years later, by fate, and by the ravings of that mad Austrian, Adolph Hitler.

Between the lives of those early Irish and myself lay several thousand years of Celtic culture and also several thousands of years for the development of an efficient agricultural science. It was those early Celtic farmers who best understood the secret power that lies in the matter we call stone. Based on that

Round tower at Rattoo, County Kerry. This tower is seven miles west of the town of Listowel and one of the best preserved in Ireland.

knowledge, Christian Irish monks built numerous elegant structures, many still intact, called round towers. The towers are fifteen or so feet in diameter and seventy to eighty feet high. They are made of stone and dot the Irish countryside.

There is no doubt in my mind that the knowledge and engineering skill that is displayed in these unique stone structures was passed down through thousands of years from the Egyptians, and that our modern understanding of certain types of ancient stone structures should take into account the possibility that there existed on this earth widespread and sophisticated channels of communication.

Professor Berry Fell of Harvard University, in his book *B.C. America*, writes of a lost America that is little known, or not even believed to exist, by present-day archaeologists. He writes of ancient Egyptian, Phoenician, and Celtic artifacts that have been found all over the North American continent. Even more astonishing, Celtic writing called Ogham script (figure 1, woodcut on page 38) has been discovered in New England and Oregon. *There are megalithic dolmans in New England!*

Dr. Fell's scholarly work sounds the death knell for the notion that the ancients did not have any contact with North America.

Over the vast eons of time the grinding, chewing, heating, and explosive forces of the drifting continents caused tremendous cosmic energies to penetrate and become trapped within the solidifying minerals of stones and clay. One of these energies is the little understood force called *paramagnetism*. If, as I believe, the ancients knew how to manipulate this subtle force, and indeed even communicated this secret knowledge among one another, then our modern knowledge of that system must surface in the collective memory of the descendants of those ancient Celts—the modern Irish.

4

THE BOOLEY PEOPLE

THE ERNE VALLEY, seen from the northern glacier-scarred moorlands called the Pullan, is beautiful. The same scene from the top of the cliffs to the south of Lough Erne, however, is breathtaking.

In days past, the view from the top of the cliffs was not easy to obtain. I remember the struggle Officer Letriem of the R.U.C. and I had climbing up to the wreck of a subsinking Catalina flying boat that crashed one stormy winter night in the deserted moorlands that crest the top of the steep, 900-foot cliffs. The aircraft had somehow missed our invisible radio range beam and hit the top of a steep hillock.

The Irish do not call their vast open ranges of scrubby heather moorlands—they correctly call them boglands or "the bog." "Bog" is more accurate than moorland as bogs are considered by geologists to precede moorlands in the sequence of plant cover.

I like the word "moorland," however, for it has the mourn-

Keenaghan Abbey graveyard and Keenaghan Lake from Magheramenagh high plain where I lived during World War II. Bogland is between the graveyard and the distant hill.

ful ring of *Wuthering Heights,* or the ring of an old Irish ghost story.

The moor, especially in the dim light of a shrouded dawn, can be a scary place, as Officer Letriem and I found out that tragic morning in 1945. We were looking for crew members that might have survived and wandered away in the black of night. The top of the great limestone escarpment is criss-crossed with steep rocky gullies. Only the thickness of the heather and whin bushes kept us from plunging into one of the limestone cracks in the darkness. We found that there were no survivors.

The highest limestone cliff overlooking Lough Erne is called Magho Cliff. Today, there is a steep winding path up a gully in the cliffs. The top of the limestone plateau which in 1944 was covered with bog and heather, now supports two large plantations of Sitka spruce. The winding hiking trail to the top of the cliff emerges between Lough Naver Forest and Blackslee Forest.

Pine plantations, or forest parks, as they are called in Ireland, are erupted all over the moorlands, mostly planted with Sitka spruce, a western American tree that tolerates damp weather and acid bog soil. One might also see parks of Norway spruce, Douglas fir, hemlock and even small plots of redwood. However, Sitka spruce does best on the high Irish boglands and exposed plateaus of western Ireland. Without the American Sitka spruce it is doubtful if Ireland could have been reforested as effectively as it has in the past two decades.

As the upland bogs disappear under the shadows of spruce plantations, one experiences a certain nostalgia for those haunting highlands where once the booley people, the cow herders from the lowland villages, spent the time between May and the end of October. Ireland, however, is in need of wood, which points up an ever recurring dilemma brought on by man's basic needs. It also exemplifies one of my own dilemmas, embodied in the word *change.*

Cardinal Newman once said, "To live is to change, and to be perfect is to change often." One of the greatest subterfuges of the modern practice of science is the belief that suppressing emotions in a scientific work renders it objective—this in the face of the fact that science is the most emotional occupation ever devised by the fertile brain of mankind. It is precisely because we pretend that science is pure and objective, and thus totally unfeeling, that we wreak havoc upon our own souls and indeed upon the very environment that supports us, body and soul. For these reasons, I do not agree with Cardinal Newman that change makes one perfect. I am certain, however, that change can most certainly develop in one a genius for survival. Ireland is an entire nation of survivors.

As I stood at the rim of Magho cliff in May of 1980 scanning the steep rocky face for the pair of peregrine falcons that each May return to nest, I could barely see the McGee cottage at the

Thatcher evens out the straw (reeds) at the eaves of a cottage. The straw is held in place by large wooden "staples" over lengths of hazel rods.

edge of the Pullan far to the west. I was looking out over a land that has survived not only the upheavals of volcanic activity, but just as certainly the impact of man's manipulation of the face of Ireland. The little McGee cottage on Keenaghan Lake is a good example of that change we moderns call progress. In the forty years between June 1944 and June 1984, the face of Ireland, especially the west, changed dramatically—for better or worse the reader will decide for himself or herself by the last chapter of this book.

The cottages of the Erne valley are typical of most of the thatched roofed rural cottages of Ireland. Many of these lovely whitewashed farmsteads have lately been drastically changed. There are very few thatched roofed cottages left in the west. By a very few remaining I do not mean that most have been demolished or replaced with modern cottages; rather, like the bog-covered high plateaus, they have had a face lift.

The boglands above the cliffs of Magho, and even parts of the Pullan, no longer present a vast rolling and smoothly covered vista of purple light—the faces of these moors have acquired the dark stubby beard of spruce. In the affairs of man practicality always supercedes esthetic considerations, whether on the moorlands of Ireland or the beaches of Florida where I live.

Exactly the opposite of the bearded effect of the moorlands has taken place on the rock-walled face of the Irish cottage. The delicate hued white limestone farmsteads are no longer visible across the vast sweep of the Irish highlands. The walls have been smoothed and stuccoed over and the light yellow thatches torn away and replaced by shingles, tile, or slate.

Despite its whiteness, the typical Irish cottage blended perfectly into the green and purple hued sweep of the open countryside. Today, the distant vista of the land has been roughened by the forests but the face of the Irish farmstead has been shaved. The walls of the modern cottages no longer are mapped with cracks or crevices along which rose and ivy creeper travel.

The visibility of the whitewashed cottages accentuated the

Poulnabrone Dolman in County Clare in the west of Ireland. The displaced farmer carved out a living on small farms in this rocky limestone region called the Burren. A dolman is an ancient megalithic monument—use unknown.

appealing nature of the countryside, whereas the more subdued stucco cottages stick out today like a wart on a fairy's nose. The change in the face of the Irish cottage indicates, at least to me, the demise of any knowledge of a special natural magic that every Irish rural family once understood. To visualize what I mean we must first understand two mysterious energies that the ancients understood, but that modern man does not at all, despite the fact that he has scientific terms for both these energies. The Chinese call them yin and yang. The Irish had no particular name for these forces but saw them in the abodes of fairies and leprechauns.

A literary and historical review seems to indicate that latterday knowledge of these energies was most evident during the hard years when England totally repressed Ireland, and, in fact, practiced genocide on the little green island. Ancient knowledge of these magic forces, of course, can be traced back to megalithic times.

All wounds are better left to heal, but the so-called Irish famine of the 1840s was in reality no famine at all. There was indeed a failure of the potato crop due to a blight, but most certainly no failure of the total food supply of Ireland. As Thomas Gallagher points out in his book, *Paddy's Lament,* Holland, Belgium, Germany, and Russia all fell victim to the very same potato blight. Unlike England, however, they ceased all food exports of grain and cattle to feed their farmers, whose main staple had become the potato. He wrote, "In every harbor in Ireland during this period, a ship sailing in with Indian maize from America passed half a dozen British ships sailing out with Irish wheat, oats or cattle." The British landlords of Ireland made a killing on the famine.

What do the hard years have to do with the subtle Irish magic that was evident until just lately? The customs and traditions of a rural agricultural society almost invariably are dependent on the geography of the land and the ecological needs of the food producers. Although geologists cringe when one says it, Ireland is in fact a bowl—a flat, central plain surrounded by hills and rugged mountains. During the hard years most of the native Irish were pushed out of the central fertile plain into the rocky circle of mountains that ring Ireland. Along the western shores and the coastal islands, from Kerry in the south to Donegal in the north, the displaced farmer carved out a living by creating new farm lands. This superhuman feat of survival, building soil where none existed, was accomplished by dragging seaweed ashore and mixing it with limestone and the volcanic grit eroded from rocky hillsides. Stones dug from the fields were either cast aside or used to fence the land with walls that the Irish to this day call "ditches."

Further inland, land was built utilizing cow manure in place of seaweed. However, the chemical principle was the same—organic matter was mixed with limestone and volcanic grit to form an extremely fertile soil. The whole methodology derived straight from the early Celtic system in which the forests of midland Ireland were slowly cleared and small patches of arable land fertilized with cattle manure and stable litter. During the summer, cattle were turned out to browse on stubby forest undergrowth. In the rocky west, instead of clearing forests the small farmer cleared away rocks, and instead of his cattle browsing on forest undergrowth, they browsed on heather. The herding of cattle to the mountains, even as late as the mid 18th century, was called booleying from booley, meaning "milking place." Booleying was an ancient herding system in the mountainous area of Europe across Spain and as far west as Switzerland. The system originally revolved around what is called the *village and booley system* where for protection farmers lived in clusters of houses (villages) and went out into the fields from a central spot to farm and herd. In western Ireland after the famine, however, instead of groups from a village spreading out over the upland in the summer to booley, the Catholic Irish, who had already spread out and were dispersed on the small plots on the lower slopes, moved to the higher mountains to booley in groups. They pooled their resources to form little assemblies of temporary shelters. It was a sort of nomad's life in which the youngest members of the farm, and mostly the girls, left the scattered lowland cottages and migrated in the

Magheranaul—Megalithic wedge-tomb near Ballyliffin far northern County Donegal. There is a strange 15 cm hole in the single massive entrance-stone of this structure. The use of such holes remains a mystery, but what is certain is that there is a subtle energy around such magic spots because cattle always prefer to graze on the grass in a wide arc around "magic" spots as shown in this photograph.

summer to the mountains, each with her family's herd. The departure was usually launched with a ceremony, and the return, at the end of October, was celebrated on Halloween. . The booleys themselves evolved into sort of mountainside villages of young people. Shelters were built of rocks and sod and the tops roofed with branches of heather and thatch.

On the mountain slopes the herders lived a rugged, healthy life among the schist and on granite-strewn slopes. There they absorbed into their bodies the very energies from those stones from the sea that the butterfly of Slieve League knew about in that unique earthy way that every insect understands. This understanding that wild creatures posses, may have suppressed by coming to depend too much on his visual and auditory senses. It is what 19th century philosophers called instinct and what 20th century mechanistic zoologists call sociobiology—the inheritance of common cultural values. It is an understanding that my scientific side calls inherited memory and my uninhibited (by modern technology) spiritual side calls a feeling for leprechauns. It is a spiritual instinct that every modern scientist has in his heart, but suppresses, and every modern farmer feels from his soil, but in modern times ignores because he has been brainwashed by modern chemical technology.

Present day religious philosophers talk about it but do not really believe, for they have come to look with 20th century arrogance on the rock-dwelling hermits and saints of old. They have exiled this ancient understanding of magic natural forces from their technological treaties with one word—*superstition.*

How did the booley people in their mountain land come to understand these weak natural forces so well? In the same way, I believe, that the ancient megalithic people and later Egyptian civilization discovered them—by using their own bodies as detectors. They sat among the rocks and felt in their very beings the same rock energies that drew the butterfly to a special mountain rock on Slieve League.

Pushed by the vicious English-imposed landlord system from their own lands, they were forced, like their megalithic and Celtic ancestors to survive in a rocky bogland wilderness. A part of the booley girls' knowledge of nature-forces lay in their inherited memory, passed on from their Celtic ancestors; but they must have developed their utilization and manipulation of such forces for practical living through the process of experimentation. This is exactly the same way that any modern high priest of science from Harvard or Yale, or at the U.S. Department of Agriculture Insect Laboratory, does it—by the scientific experimental method. The latter are called scientists but booley girl "professors" were labeled witches and burned at the stake in every country of the world except *Ireland*!

By the time I arrived in Ireland in 1944 there were no true booley people left. The system of herding cattle to the mountainside had long before been discarded and sheep had become the grazers of the hillside moorlands. Nevertheless, the knowledge of the fairy and leprechaun forces was dying hard, and on more than one occasion, in my rambles across mountain and moorland, I met up with Irish herders or moorland cottage people who understood perfectly the forces in the rocks, the turf and bogs, and the singing moorland winds.

On the back slopes of the massive Magho Cliffs, approached from Parrancassidy Cross Roads to the south of Belleek, I often stopped to talk to cotters and herders. Later I met the same solitary types on the steep northern slopes of Slieve League or the south-facing slopes of Ben Bulbin above the beautiful valley of Glenade—a favorite magic spot of William Butler Yeats. Although they no longer camped in wattle huts among the singing winds, as the booley girls did, they nevertheless spent many a spring and summer day in solitude among those magic hills. There was plenty of time for them to sit quietly among the rocks and bathe in that energy just as I did sitting outside the door of my lonely outpost radio station at Magheramena Castle.

I could gaze south across that magic land to the cliffs of Lough Erne, or north to the rock strewn slopes of Breesy Hill where I eventually found my own magic stone.

I first stumbled across that stone while I was following a little accepter hawk (called a sparrow hawk in Britain) through

My magic (paramagnetic) rock on top of Breesy Hill near Belleek. The view is north across the vast Pullan.

a grove of trees near the Mile's cottage at the foot of Breesy Hill. The sparrow hawk is still common in Ireland but like the American yellow-billed cuckoo it is seldom seen because it is a secretive little raptor—a denizen of the leafy tree tops.

After I lost sight of my hawk I started up the side of Breesy Hill. It is rock strewn but not steep, and at the south edge I stopped to rest. I sat at the base of a large boulder with a flattened side. The cone shaped mountain is called Breesy for good reasons: A gentle wind blows night and day across the rocky top.

On that long ago day I soon began to feel as I did as a youth when I looked in my little sister's eyes after telling her a fairy story I invented, or when I watch my wife at her gardening, or when I first saw my paratrooper brother alive in Ireland after he was wounded in the Battle of the Bulge. It is a strange and wonderful feeling which saints must experience when they think about Christ or mothers after giving birth. I suppose the best term for such feelings is bliss or tranquil happiness. It can be temporarily gained, of course, with no effort whatsoever by taking drugs, tranquilizers or marijuana, etc., a stupid and fleeting way to gain a worldly joy that soon becomes incoherent and frustrating, and, worse yet, slowly but surely poisons the body and warps the brain.

I am sure I first discovered joy in plants and rocks in Ireland, although I certainly suspected it climbing among the cliffs at Red Rocks near Denver and earlier as a child picknicking at that magic pile of desert rocks near El Paso called Hueco Tanks. It was at those magic places that I dreamed as the booley people must have dreamed. I understood, without knowing why, how the Irish could have survived through centuries of hardship and warfare and still maintain a certain outward tranquility that few people anywhere in the world can match.

What William Butler Yeats expressed poetically I shall attempt to put into ordinary words, based on modern concepts of science—words that everyday people, the common sense people of the world, will understand.

I now know that the weak magnetic forces that modern science calls diamagnetism and paramagnetism are the magic forces the ancient megalithic and Celtic peoples found in stone and plants.

I have depended a lot in my rucksack journeying on "kitchen" experimentation and common sense, but I also credit my inherited memory, which was passed on to me through generation after generation of Irish and Spanish ancestors. That memory was fired anew when, like the butterfly of butterfly rock, I sat still and quiet at the foot of my own magic rock on Breesy Hill long, long ago. In those days I did not even know that there were booley people; however, my inherited memory probably knew. One of my ancestors, I do not doubt, was a moorland booley girl and she walked among those Irish Hills humming a gentle melody to accompany the singing moorland winds.

5

THRAW-HOOK AND SLANE

MOST MEMORIES DIM with time; however, some instinctive beliefs are etched as firmly into the circuitry and storage bank of brainmatter as are the hieroglyphics on the limestone walls of the great temple of Karnak at Luxor in Egypt. Perhaps it is, as some Irish believe, only the second son of the second son whose memory inherits and reprocesses such instincts. If so, at least in my case, it will have to be the second son of my first daughter whose memory encompasses my own instinct for *creature magic*. Strangely enough he was born in Ireland, whether by coincidence or design only the Creator knows.

By creature magic I mean that magical empathy that certain quiet people of the world have with life forms other than human beings. In my case, since my earliest youth it has been for birds. If a learned Christian theologian were to argue with me that birds do not have souls, I would argue back that the learned Christian theologian was a learned Christian fool.

The Indians at one time knew all about creature magic, but

now, unfortunately, have mostly lost it. Henry David Thoreau certainly understood it for he once wrote:

> This especially is the bird of the river,
> There is sympathy between its sluggish flight,
> and the sluggish flow of the stream—
> Its slowly lapsing flight,
> Even like the rills of Musketwork
> And my own pulse sometimes.
> ["All Nature is my Bride" ed. William M. White]

The pulse is the wave of life. It ebbs and flows as does the ever recurring flow from fear to love that governs all life. Fear is the servant of the unknown, love the servant of God. To fear is to doubt, to be in the state of love is to be God-ful. If scientists and preachers could understand this, there would be neither nuclear bombs nor fire and brimstone theology. If such intellectuals truly understood, and had inherited any memories of creature magic, they would not build such bombs or preach such sermons.

I have always believed in creature magic. The tales of Albert Payson Terhune about the marvelous dog Lad probably reinforced my belief in the magic of bird and beast, but so also did the gentle nuns at Blessed Sacrament Grammar School in Denver. Those nuns were indeed gentle, for I was definitely loved and taught, but never once struck by a nun, modern disciplinary mythology to the contrary.

It was my mother who taught me from the catechism that "man is made in the image and likeness of God," and that "God is everywhere." If that is so, I reasoned at a very early age, then God is so big and so all encompassing that surely my dog and every singing bird, and even little ants, must be made to the image of a little tiny part of God. Apparently I have never grown up for I still feel this is so. If mankind is the mirror image of God, then my parrot Kluck, who sits with patience on my shoulder, is the mirror image of a little fleeting-floating part of God.

It was the creature magic part of my boyhood memories that

The ruins of the two roomed Irish cottage where the great blue tit flushed miller moths from the thatch with a straw "tool." The owners McGarrigal are long since dead and the little cottage deteriorates in ghostly silence.

caused me to pause and smile, and to blink my eyes in astonishment, when, on the upper lane to the Ghosta river, I, not so long ago, spotted the black-bibbed great tit hard at work on a cottage thatch.

The Ghosta river is, in truth, not a river, but a little creek that separates Northern Ireland from Southern Ireland near Belleek. It flows from Keenaghan Lake and crosses the back lane that takes one from Belleek road in the north of Ireland to the McGee cottage in the south. The upper lane passes by a thatched roofed cottage that during the war was inhabited by an old and friendly Irish couple.

Often as I crossed the border along the back lane into neutral Ireland, I was greeted by the woof of their dog and the soft murmur, like a flowing brook at eventide, of the old couple saying their rosary before the open hearth. These were welcome and comforting sounds on a dark and lonely night crossing the border to the McGee house.

Years have passed and today the little two-roomed cottage is in ruins. The limestone grey walls are crumbling, the windows and door hinges broken, and the thatch has a crown of green weeds growing from its once trim edges.

The sprightly, black-bibbed great tit that I stopped to watch was hanging from the edge of the thatch. It fluttered along the ragged eves, stopping now and again to pull a straw from the thatch. As it hung from the edge of the decaying roof it cocked its head to the side as if listening to some far away voice from deep within the yellow straw. It was listening for the scratching rumble of a moth or fly larvae deep within the dried weeds. I knew that the sound is a scratching rumble, for my friend and former colleague Joe Benner once let me listen to the sound of a fruit fly larvae deep within an orange. Joe worked in our laboratory on sound experiments. Man requires a very sensitive crystal microphone and a special noise-free acoustic room to hear the low rumbles, but the black-bibbed great tit needs no such device.

On that morning the tit paid not the least attention to me for creature magic was at work and he had no fear that I would disrupt his attention to the job at hand. He cocked his head and poked and poked with his tiny pointed beak, all to no avail.

Finally he hopped to one corner of the scraggly thatch and grasped a long straw in his beak. He stretched his legs out as far as was possible, and with his beak tucked down against his black breast, reared back, straining with his outstretched legs until it appeared that surely they would become detached from his feathered body. As he pulled he sawed the straw back and forth. As the straw pulled loose and fell to the ground, a sleepy brown miller moth, disrupted from its hidden crevice by the movement of the straw, fluttered to the surface of the thatch. The bird snapped up the moth before it could recover its wits.

A second moth emerged and launched into flight but the tit grabbed it and flew to the top of the stone wall where he devoured his fluttering morsel. Creature magic had allowed me to witness what I thought was the first observed utilization of a bird chop-stick. [I found the same behavior described in an old series of British papers called "Birds of our Country"]

To use a tool, straw or otherwise, is one criterion for higher reason in mankind, but a scientist would surely clothe the description of such an event, in the case of a bird, in behavioral jargon to prove that it is a mechanistic phenomenon with no learning or reason involved. Elitist scientists no longer feel that little fleeting-floating part of God, put into their hearts as children.

Who can really tell, scientist or not, where inheritance, or instinct if one prefers, leaves off and reason and learning begin? Certainly, despite their technical jargon, no scientist, including myself, that I ever met.

If between living creatures and certain people there is an inherited empathy that I prefer to call magic, so also might there be between the typical whitewashed Irish cottage and those who dwell therein. In fact, I never met a single person, whether or not he lived in such a cottage, who did not agree that such a dwelling is elegant beyond all description. Poets dream and write about them, and tourists exclaim at the very sight of such a home. Today the Irish, who now call in builders to tear away the thatch and transform the uneven rock walls, exclaim "aw shure the thatch is no good atal," at the same time knowing in their hearts that they have lost a little piece of their soul. They will not admit it, but it can be seen in the eye of each and every one.

The strange look in the eyes of the cottor, whenever one mentions the old home, tells me that just as there is a creature magic, so also there must be a cottage magic which is slowly disappearing from Ireland, as it did earlier from England. Any kind of magic, of course, whether creature magic or cottage magic, must depend on some kind of subtle God created energy. In the case of creature magic it is without doubt mediated by the aura, as mystics and scientists alike call it, but can a cottage have an aura? Of course it can, and I shall prove it in this book, utilizing the scientist's own weapon of experimentation. I shall draw a picture of the aura of the limestone, thatched roofed Irish cottage. It is one of the most peaceful and soothing auras of all of the dwellings ever devised by mankind. First, however, let us trace the ancient origin of the Irish cottage.

The origin is very important for its source was in the hardship of eking a living from the rocky, windswept land that is Ireland. It involves also the constant warring to control the bounty of that land and, strangely enough, the cold, singing winds that sweep across the rugged Irish hills and mountains.

A rusty old thraw-hook from the McGee cottage hangs on my den wall. The thraw-hook is the basic symbol, for me at least, of the old way of life that is disappearing from Ireland. It is also the farm instrument that holds, so to speak, the sound of the winds in its twisted form. One can almost hear the sighing breezes off the Pullan moorland in the thraw-hook on my wall.

Though the thraw-hook is a simple instrument, it was once as important to the small Irish farm as the donkey and cart of later years. There are several different types of thraw-hooks but most followed two basic patterns. The common type, and the

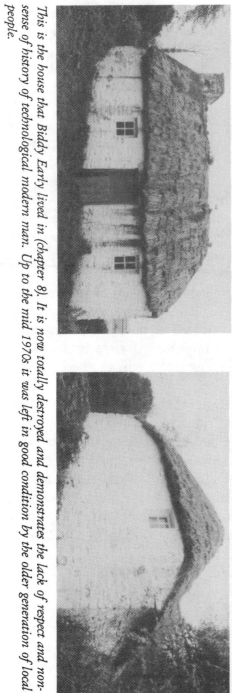

This is the house that Biddy Early lived in (chapter 8). It is now totally destroyed and demonstrates the lack of respect and non-sense of history of technological modern man. Up to the mid 1970s it was left in good condition by the older generation of local people.

one almost always found in the north where I lived, consists of a strong wire crank with a handle at one end. The long straight part of the wire is inserted through another handle which has a hole in it so the heavy wire can rotate. The type used further south in Kerry and on Dingle peninsula, was merely a stick bent at right angles with a T handle at the opposite end. The Donegal type is the more modern version and easier to use.

The purpose of the thraw-hook is to twist straw rope out of oat, rye, or hay cut from pasture lands. Twisted straw rope is used to tie down the ricks of hay, and also, at least along the Donegal coast, the straw thatch of the cottages against the strong fall and winter winds that sweep from the Atlantic Ocean.

The drying of hay for winter cattle feed probably began in Scandinavia and reached Ireland in later centuries by way of England. In Ireland it is called "savin" the hay, which in essence means drying it.

In order to twist rope, handfuls of tangled hay are pulled from the small cocks (piles) by one individual and fed to the end of the thraw-hook which is twisted by another. The hay rope for tying down is thus manufactured on the spot.

One might think that the Irish would curse those sweeping winds, but without them, neither the thatch of a roof nor the ricks of straw would ever dry. It is those sweeping winds that save both the hay and the roof.

The placement of the two-roomed Irish cottage was also dependent on the sweep of the winds across the landscape. In most cases it is possible to date a cottage by its position against the hillside. Cottages built before or during the bad years (the 19th century famine) were most often placed with their gabled end or back directly against a cut in the slope of a hill—stuck into the side of the hill, so to speak. In later years cottages were placed with the long axis of the cottage horizontal to a hill, but never cut into the hillside. An excellent example of the former early type is the famous Irish healer Biddy Early's cottage, built by her husband in the middle of the famine years (1840s).

Biddy's cottage is fascinating, for it is a hybrid of two standard types of cottages (see four photographs). In the north, the typical cottage most often has a stone gabled roof with the fireplace at one end. In the south, the thatch slopes over the end of the cottage and the fireplace is in the middle between the two rooms.

I say that Biddy's house is a hybrid because the chimney end is gabled and inserted into the hillside, while the opposite end is sloped over with thatch, called a "hip" roofed thatch.

The McGee cottage across the northern border in the Donegal townland of Bonahill sits horizontal to the contour of the hillside, and though it has stone gables at either end, the chimney is in the middle like the hip roofed houses of the south. That is because after the hard years and the end of the landlordism in Ireland, when the small Irish farmer attained ownership and prospered, he often added a third room behind the fireplace, so that the chimney ended up slightly to the side of mid-roof. The McGee cottage was expanded about sixty years ago (photo on page 79).

Although the back of the McGee house is cut into the hillside horizontally instead of gable endwise, the very fact that it takes advantage of the insulative properties of mother earth along its rear, and is sheltered from the Pullan winds, indicates a more ancient construction. The cottage is known by the locals to be at least 400 years old and perhaps even older.

Very few post-famine houses are cut into a hillside, and today hillsides are leveled and modern cottages built in the center of the flat space, open to the cold wind, or raised above the slope, and thus less insulated. It is difficult nowadays to tell whether lack of common sense or contempt for the old ways is responsible for this absence of judgment—probably the latter, for the modern Irish still exhibit an above average degree of common sense.

The paradox, of course, is that modern Western man seems to be solving his problems with words rather than actions, thus, although he has a pretty name for the miserableness of cold air mixed with wind (the wind chill factor), he now builds his home, not to shelter himself from wind chill, but rather to

The McGee cottage, Bonahill, southern Ireland. The Callahan and McGee cousins are running across a field a few yards from the North Ireland border. Hay is put up in ricks to dry. Later they are built into larger stacks (ricks) and tied down with straw rope woven by a thraw-hook.

amplify it by the shape and materials used in his modern house. Obviously the colder the walls the more money the local electric company makes to heat the house.

When the bottom half of a cottage is buried in a hillside only the top half is exposed to the chilling wind. The lower walls are protected, and since hot air rises, only the exposed top half of the wall steals warmth, and that is the half where rising surplus heat is available from the hearth.

Whether half buried in a hill or not, there is another very fundamental difference between the hip roofed cottages of the south and the northern stone gabled cottages. In the gabled house the byre (cow barn) is usually attached to the stone gabled and opposite the bedroom. Since the cows in a small byre generate a lot of heat, the outer bedroom wall is protected against cold and wind. The byre of hip roofed cottages is usually separate, but the thick thatch curves down over both ends of the cottage and protects them. By any standard, thatch is the best form of insulation ever devised by man. It is to a house what a down cover is to a bed. Since the hearth in a hip roofed house is in the middle, with one room on each side, it is more evenly heated than a gabled house with the kitchen hearth against the far wall.

Another peculiarity of the Donegal cottage is that it often has an outshot (long notch) along the back wall next to the fireplace (figure 3, page 82). A curtain usually hangs across the face of the outshot. In olden days a bed neatly filled the outshot, and a child slept there close by the banked fire. It was also a good place for a sick or older person. Today, with most cottages expanded like the McGee cottage to three rooms, with an addition behind the hearth, the outshot is used for a couch or for storage.

Turf, called peat in Southern Ireland, is still burned in Ireland though seldom in an open hearth anymore. Turf is extremely slow-burning and clean. It has a marvelous characteristic smell.

As important as the thraw-hook is for tying against the wind, so is the turf slane (spade) for cutting the turf. There are

The O'Shea and Callahan young people bringing in the turf. The Magho Hills where the Catalina RAF aircraft crashed is in the background.

almost as many types of slanes in Ireland as there are regions, but generally they all fit one basic pattern. I have a Donegal type. The spade end is a foot long and four inches wide. The edge of the slane has a right angle flange, of triangular shape, that sticks out five inches. With this specialized turf spade, turf bricks about ten by four inches are cut from the damp bog. They, of course, will not burn when wet since turf is essentially moist, but partially petrified, plant life (pre-coal stage of development).

The turf is stacked on the bog in wedge-shaped piles and the same wind so necessary for saving the hay, also dries the turf. That is why I say my thraw-hook and slane have the winds in their rusty bodies.

It is a paradox that the cold moorland wind is so necessary to the Irish cottage for day-to-day warmth. It may chill the body but it dries the turf for the fire and dries the straw for the thick insulative thatch.

Another characteristic of the Irish cottage is the thick walls (two-and-a-half feet across) that are not solid rock, as in modern concrete brick houses, but are double-walled and filled between

Figure 3. An Irish gable ended two roomed thatched cottage (top). The floor plan shows a northern cottage with an outshot and the byre at one end (center). Such cottages are often modified by the addition of a room behind the fireplace with an entrance door beside the fireplace. New thatch is usually laid over old thatch. Rubble fills the space between stone walls (bottom).

John McGee with the turf spade (slane).

A flanged Irish turf slane (spade) showing back (right) and front (left). The slanes from different sections of Ireland take slightly different shapes. The long flanged form is typical of County Galway.

with rock rubble. The rocks form millions of small air pockets—the same principle as styrofoam insulation, only better.

The genius of our Irish ancestors was that they could design an inexpensive, yet highly efficient, dwelling that takes advantage of so many laws of physics, without knowing any physics. (Or did they?)

Perhaps in building their insulated wind-cured cottages they instinctively knew more physics than we, with all our ten dollar words, and stupidly designed, power wasting houses ever will know.

It is indeed sad to see such highly efficient dwellings disappearing from the Irish countryside, not just because of their beauty and physical efficiency, but because the Irish farm cottage was designed to generate that same unique magic force that certain shapes of rocks generate naturally and that I could feel on Breesy Hill. Such a cottage, properly placed, encompasses its inhabitants with an aura of serenity and well-being felt in no other dwelling.

Once the thatches go, where will the black-bibbed great tit find little brown miller moths to pull forth with his straw chop-stick? Will the great tit mourn the loss of the white-walled, thatch-roofed Irish cottage as does this rucksack naturalist?

6

THE HEDGEROW ZOO

A HAZY BLUE-GREEN LIGHT filled the valley of the Erne. From the top of Magho Cliffs a panorama of thin darkened lines spread like a huge fish net across the lush valley. These were the hedgerow fences of green that separate each little farm.

In the distance the ether-like pattern appears as a checker-board, but close up it is so irregular and topsy-turvy as to bewilder a knight or pawn on such a playing field. That is because the thin dark lines are barriers of whitethorn tree and grey rocks that were set down unplanned by any modern parking lot corporation. The parking lot corporations will eventually take over, however, and the topsy-turvy playing field will revert to one huge flattened expanse of grass.

I could imagine a nice silver chain link fence along the Bonahill road from Kennaghan lough to Ballyshannon along the western Pullan side of the river, and an equally "lovely" fence from the base of Magho Cliff to the edge of Donegal Bay. The long fences would confine a vast herd of 10,000 cattle. One edge of Ballyshannon might become the modern booley spot—called a feed lot. The pollution and manure from Ballyshannon booley-feed lot would then be carted eastward and dumped in Lough Erne.

Eventually the checkerboard fields would no longer be owned by a thousand small farm families but by a group of modern powerful lords and knights, called a Board of Directors, issuing decrees, called "efficiency" reports, from giant glass, metal and exceeding ugly castles, called skyscrapers, in Dublin. This modern scenario of the feudalism of the Erne valley is not as exaggerated as it might appear in these paragraphs. It is well on the way to happening in the British Isles.

Corporate feudalism is being subtly propagandized as an efficient system for the small farmers in Ireland, as it is all over the world. The freeholder, as the Romans called the small farmer landowner, will soon be forced from his land by the feudal lords of agribusiness.

At the very time that the Irish have thrown off, by revolution or negotiation, the power of their Anglo-Saxon landlords, and I include Protestant and Catholic (North and South) alike, they are being talked and taxed into believing that they are lazy and inefficient farmers and told to move to the city where they can get nice jobs in industry—jobs that do not and cannot exist in a land where the only sufficient resources are agriculture and beauty (tourism).

The paradox of this 20th century transition period is that the development of landlordism, through the period historians call the age of enclosures, took centuries, whereas the reversal to corporate feudalism will encompass a very few generations.

From the top of Magho cliffs I could not see the O'Shea cottage. It was lost in the blue-green misty vapors. I had spent many of my happiest hours visiting in the O'Shea cottage.

In days past I often stood at the door of the thatched cottage swallow-watching. In the waking dawn, after the dew was burned from the straw by the rising sun, the swallows dived and swooped upward to grab the gnats emerging from the edge of thatch—usually in total oblivion to my frame blocking the low doorway. I have had many bumps from entering an Irish cottage.

Sometimes a swallow would fly past my ear directly into the kitchen, swoop by the huge open hearth and out again into the cool dawn. It was creature magic at work. Two little swallow families nested in the byre where they kept an eye on the cows and milking stool.

It was in that byre that I witnessed my first birth. Anthony pulled the little calf from its mother while I held the lantern. There was no electricity in the Belleek countryside then. The O'Shea cottage was typically end-gabled, with the huge open hearth at one end and the byre by the opposite gable. Behind the hearth was a bedroom.

Anthony's father was a retired schoolmaster and, like my farmer friend John Gormley, who lived up by the castle grounds, was more than willing to sit by the hour talking about ancient customs, such as the planting of a rowan tree by the Irish, or highland croft.

The word croft is of Scottish origin and usually is applied to the small enclosed field nearest to the house. It is not unusual to see a lone rowan tree planted next to the croft nearest the byre where the milking cows or newborn calves are turned out in the spring.

Rowan is the common name given to a beautiful white blossomed tree, *Sorbus aucuparia*, sometimes called mountain ash because it grows right up the sides of mountains. It is not, however, even closely related to the ash family of trees.

The Gaelic word for the rowan is caorunn and the Welsh, cerddin. Rowan is probably of Danish origin from "ron" or "rune" which translates to "magic tree." It is the tree that guards best against witchcraft. Because it is a tree that needs light it is most often found along hedgerows rather than in the forest. If a large and ancient tree sits near a byre or croft one may be sure that the house itself is of ancient origin for the tree was originally planted to guard the cattle. The rowan seldom exceeds 35 or 40 feet in height and the beautiful spring blossoms are white with an indescribable, not very pleasant odor.

The whitethorn after leaf fall. A few white blossoms can still be seen (lower left). The magic whitethorn, called hawthorn in England, is hung with large white haws (berries) in the fall.

Unlike the hawthorn tree, which in Ireland is usually called the whitethorn, it does not have thorns and therefore is seldom used to form hedges.

The scrublike whitethorn, *Crataegus monogyna,* is one of the commonest deciduous trees in Ireland. Its thick thorny branches form the main body of the most workable hedgerows. Like the rowan, the whitethorn has its own special place in the so-called superstitions of country folk—it is considered a fairy tree and therefore brings bad luck to anyone why cuts it down. Efficient farmers, however, keep the tops well pruned to encourage heavy, impenetrable lower growth. It is impossible for man or beast to get across a properly maintained hedgerow. One must seek the gate or stone style to cross from Irish field to Irish field.

To understand the importance of the hedgerows and their place in the farming methods of Ireland, or southern England and the European mainland, one must view the hedges against the history of the enclosure of the wide expanses of open countryside, formed after the last of the forests gave way to agriculture. The enclosure of land with hedges changed many things, including the sport of falconry which was difficult to practice without wide open spaces.

There are three main types of what, in Ireland and England, are classified as hedgerows, though the Irish call them "ditches." One, of course, is the solid stone fence which is bare and contains no green growth. The second is the tree hedge composed mostly of whitethorn trees but which may include other deciduous trees, such as ash, beech, elm, or sycamore, and smaller, more scrubby bushes such as hazel, wild rose, blackthorn, and sour cherry, plus, of course, numerous small cover plants of ivy, oatgrass, brambles, nettle, hogweed, herb robert, and privet.

In England where this type of plant growth is common, the hedges are rebuilt every ten to fifteen years. The laying of hedges is still practiced in the southwest of England, especially around Dartmoor and the lower border of Wales where thick hedges are needed to confine stock. In the west of England they

An ancient unkempt whitethorn hedge planted on a typical earth covered rock mound in olden times. As the whitethorn grew old and was not trimmed it grew into trees. The gaps are now filled with modern wire fence.

have hedge-laying contests.

The process consists of partially pruning each hawthorn and cutting half through the trunk a few inches above the base of the tree. Each tree is bent down and laid over on top of its neighbor. To keep the supple thorn trees from springing back up, stakes are pounded into the ground every few yards and flexible hazel rods woven in and out along the tops of the stakes. The woven line of hazel keeps the pushed-down hawthorn from slowly straightening up. New growth quickly sprouts from the base, between the bent, overlapping trees. This grows into a thick new hedge. The laying of hedges is an ancient and honorable chore of the farmer on enclosed lands.

The hedgerows enable a farmer to move cattle or sheep from field to field, grazing and manuring one while a second remains fallow (untilled, or turned over and unseeded) and a third planted to oats, or for hay or other arable crops (wheat or horticultural plants). It is an efficient and practical rotation of crops that keeps the land sweet and fertile and the farmer busy and contented—he buys no wire or fertilizer and thus owes no bank for the keep of his land or cattle. A newly laid hedge is a sort of developing fence barrier that confines the stock while the new hedgerow is regenerating itself.

The third type of hedgerow is more common in Ireland than England, except, again, in southwest England where it is called a Devon hedge. I call it a rock mound hedgerow (figure 4, page 93). It most probably developed mainly in the rocky and mountainous regions of England and Ireland, in those areas where rocks were dug from the fields and stacked around the edges.

This type of hedge evolved when field rocks were covered with a mound of dirt, usually lifted from the ditch along the road or pathway edge. Hawthorn was planted along the ridge of the mound on the ditch side and the tops pruned off for heavy lower growth. Sometimes a few rocks were exposed but mostly they were completely covered by earth.

Hedge plants seldom grow into the ditch, which makes the cleaning and reshaping of the roadway ditch quite simple. Such

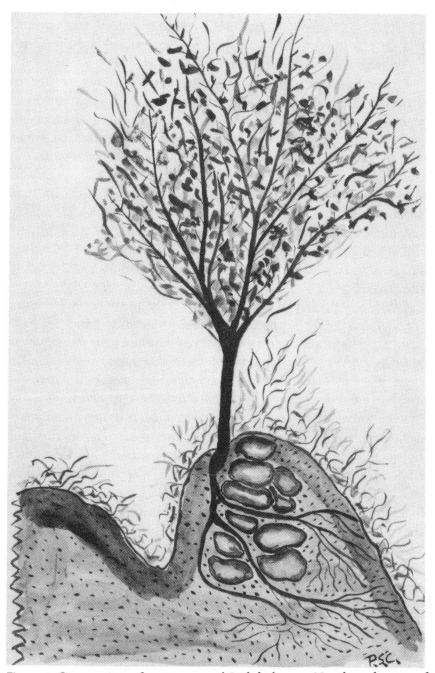

Figure 4. Cross sections of a stone mound Irish hedgerow. Note how the roots of the whitethorn tree grow towards the paramagnetic stone.

a hedgerow is easily maintained by pruning alone and is rarely laid, as the lower part is plugged by the mound and the scrubby growth that lines the top of the mound.

History books used in schools are almost always written by professors who sit at desks and interpret history in terms of war. In any college text 900 pages will be about battles and politics and one paragraph about the agriculture and daily lives of the peasant. In all such treatises, however, it is correctly stated that the Irish stone-hedge barrier, with little doubt, developed from the early Irish ring forts, stone or dirt mounds surrounding a group of wattle (mud and stick) huts. The remains of ring forts are spotted all over Ireland and represent the visual remains of the early Irish pre-Celtic defensive village. In cross section, the profile of the ring fort does indeed mimic the cross section of a hedgerow ditch (figure 4, page 93).

Originally such forts were stone walls and were well suited for defense against physical attack—especially hand to hand combat. They were probably far more important, however, as places to shelter stock at night from marauder wolves rather than marauder men. If the ancients and medievals spent as much time fighting as the history books say, chiefs and followers, lords and serfs alike would have been living in a state of perpetual famine. More than likely, due to low population and lack of good communications, their wars occurred far less often than ours, which occur one per generation (five major conflicts since 1900). Furthermore, there were probably fewer MacDonald's hamburger type shoot outs since people close to the land produce fewer crazies than do highly populated cities. Why?

We assume that the ancient megalithic and Celtic peoples were savages because they threw spears and arrows at each other instead of heat seeking missiles, flame throwers, and atomic bombs. At least ancient warfare demanded a certain amount of personal physical courage. Modern bombing is the most cowardly form of warfare ever devised.

As for being ignorant and superstitious even a college student, with a single course on the *theory* of evolution, might

There is a large concentration of ancient stone monuments south of Limerick at Lough Gur. This is the large continuous stone ring on the Bruff-Limerick road. Cows love to graze such places. Large banks or mounds around some stone circles indicate to me that many were also ring forts.

The Hedgerow Zoo 95

point out that the human brain has not really evolved over the past 5,000 years. The Great Pyramid at Giza should tell us something. That masterpiece of engineering was built about the same time as the early Irish megalithic civilizations (2600 B.C.).

In my opinion, the hedgerow developed from stone rings as a modification to keep that much maligned animal, the wolf, away from the cows and sheep. A thick step up or a shoulder-high wall is ideal for spear throwing or archery; a wall with a hedge on top of it will so obstruct vision that an enemy could approach with ease. It is impossible to shoot an arrow through a hedgerow. On the other hand, any thick thorny barrier that keeps cows and sheep in keeps wolves out. Furthermore, it can be easily patrolled by dogs, which can sense prowling wolves through the brush.

The hatred of the medievals and Celts for wolves stems from their pack attacks on stock. Wolves have evolved to kill large mammals. In North America their favored prey are elk and moose. The simple fact is that despite folk tales to the contrary notwithstanding, there is not a single verifiable attack by wolf packs on a human being. This is well documented; the Canadian wildlife authorities, for example, point out that there has never been an attack among the thousands of children who camp every year in Algonquin Park, the outdoor recreational area with the largest densities of wild wolf packs in the world. The one documented attack in Canada took place in 1942 and was by a rabid wolf. Indeed, if St. Francis really tamed a wolf that was menacing a town in Italy, then the real miracle is not the taming of the wolf—rabid animals often appear friendly—but that he escaped death from rabies.

The hatred for large predators by a people that depend on stock for their livelihood has not abated in modern days.

Judged against human behavior, and humans also kill and eat cows and sheep, the wolf is an exemplary creature. Brave, loyal, and kind to its young and mate, it never murders its own kind. Be that as it may, the disappearance of the woodland environment and the invention of longbow and gun caused the extinction of wolves in England by the late 1400s (thirteenth cen-

TABLE II
Nesting Hedgerow Birds (Moore et. al., 1967)*

Unclipped Stockproof Hedge—2,645 yards long Species (9)	Pairs
whitethroat	7
yellow-hammer	2
song thrush	2
blackbird	9
linnet	9
hedge sparrow	6
chaffinch	3
robin	1
great tit	1
Total	40

15.1 per 1,000 yards of hedge

Stockproof Hedge with Outgrowths—2,099 yards long Species (19)	Pairs
red-legged partridge	1
corn bunting	1
reed bunting	5
white throat	6
yellow-hammer	6
house sparrow	2
song thrush	6
blackbird	11
linnet	6
bull finch	2
hedge sparrow	5
wood pigeon	6
chaffinch	3
goldfinch	1
blackcap	1
garden warbler	1
lesser whitethroat	3
turtle dove	2
wren	3
Total	71

33.9 per 1,000 yards of hedge

*Moore, N.W., M.D. Hooper and B.N.K. Davis. Hedges I. Introduction and reconnaissance studies. Jour. Appl. Eco, 4:201-220. 1967. —an English study done on a farm in Huntingdonshire. Since most of these species also occur in Ireland, and since most Irish hedges are stockproof, an Irish survey would in all probability yield similar results.

tury). At that time, however, Ireland was still covered by considerable impenetrable forestlands. The last wolf was killed there toward the end of the 1700s, well within the time of the village and booley system.

As the forests of Ireland were cut down for additional grazing land, charcoal, and to build the English Navy, the wolf packs were forced to range farther and farther into the upland boglands for prey. If they were such a threat to human life it is doubtful whether the booley lads and lassies would have been allowed to dwell alone in the wilderness, with no hedgerow protection. Those children were the protectors of cows from wolves.

As late as 1584, long after wolves had disappeared from England, a certain Robert Legge wrote a *Book of Information*, which recommended that:

> . . . from destruction of ravening and devouring Wolves, some order might be had, as when any lease is granted, to put in some clause that the tenant endeavours himself to spoil and kill Wolves with traps, snares, or such devices as he may devise.
> —*in "The Irish Beast Book," J.S. Fairley*

Fifteen fifty-four is the middle period of landlordism, so the rent paying Irish tenant, already taxed to poverty, was also made responsible for wolf control.

The contempt of the landlord for the Irishman and the wolf is well stated in a latin verse by Fynes Moryson, secretary to Lord Deputy Mountjoy (1599 to 1603). Translated it reads:

> For four vile beasts Ireland hath no fence:
> Their bodies lice, their house rats possess;
> Most-wicked priests govern their conscience.
> And raving wolves waste their fields no less.

I can't say much good about lice and rats, but for gentleness, among animals, wolves have few equals. As for priests, they are trained to supplement the Catholic conscience, and indeed, human to human, if my experience is any criterion, few kindlier or more gentle people exist. Wolves and priests, however, do eat cows, as I suppose did Lord Deputy Mountjoy! At any rate, during the days of land enclosure Ireland did indeed develop an excellent barrier against wolves. It also developed a combination wind erosion barrier and hedgerow zoo. Although wolves disappeared long ago, wind, insects, and birds have not.

A list of the beneficial insectivorous birds that utilize the hedgerow for nesting, food, and shelter is given in Table II, page 97. It will be noticed that unclipped stock-proof hedges have the highest number of nesting birds per 1,000 yards of hedgerow, except for overgrown hedges with outgrowth of blackthorn, etc. Even clipped hedgerows, however, shelter a goodly number of insectivorous birds. These are the very species that feed their young almost entirely on insects during the spring and summer when crops likely to be attacked are growing and maturing—a time when they are most susceptible to insect attacks. An excellent example of how diversity makes for natural stability in nature is the fact that, in hedgerows that have outgrowth close to the ground, the number of different species rises from nine to nineteen, and the number of individual nesting birds from 15.1 to 33.9 per thousand yards.

Bees abound in hedgerows as do beneficial wasps.

Just as important as the bird species that suppress insect outbreaks are, of course, the many insect predators and parasites that help control outbreaks of insect infestations. Hedgerows also provide a habitat for many different species of lower invertebrates which in turn feed on fungi, dead wood, and the leaves of flowering hedgerow plants. Without the recycling efforts of such invertebrates, the countryside would soon be buried under a layer of undigested cellulose.

Besides providing a habitat for beneficial parasites and predators, the hedgerow is a primary source of nectar for the honeybee. Eighty percent of all agricultural crops are pollinated by the honeybee.

As I looked out over the Erne valley from Magho Cliffs my thoughts went back to yesteryears when I would leave the barracks at early sunrise and walk to Belleek. I called it my stoat and hedgehog watching walk. The hedgehog is well-named, for the little "brambley" coated animal is more common in the hedges of Ireland than is realized. Not only is it insectivorous, and thus the farmer's friend, but it also feeds on the roots of the annoying weed, plantain. Gilbert White in his masterpiece of nature writing, *Natural History of Selbourne*, describes how the hedgehog bores beneath the plant with its long upper mandible and eats the roots upwards, leaving the rest of the plant untouched. It is not at all uncommon to see such withering plantain along the banks of hedgerows. It is a sure sign that a hedgehog is close by.

I once kept a pet hedgehog for a month or so. When it was disturbed by someone it did not know it curled up into a tight ball and no amount of prying could force it open.

The Irish stoat, a member of the weasel family with a red-brown coat, lives mainly in the rocks and cover of hedgerows. It is a subspecies of the Arctic stoat, and is found only in Ireland and the Isle of Man. It is as Irish as St. Patrick, with a life history almost as interesting.

Irish-stoat-watching is best accomplished sitting quietly along a roadside ditch and sighting along the hedgerow. With patience, a whithred, or whitterick (from "white throat"), as

The ferret is a close relative of the Irish stoat and is used in Ireland to hunt rabbits. This is a trained ferret.

the Irish stoat is known in the rural north of Ireland, will stick its little head out from the hedgerow brush and dash along the field edge to some other cover.

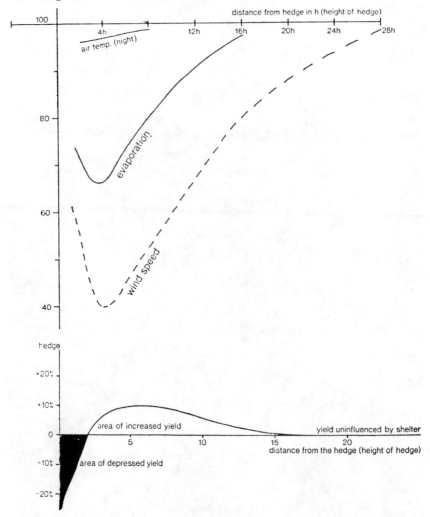

Figure 5: Water evaporation from soil decreases with a decrease in wind speed (top curves). Distance from the hedge to the center of a field is measured in multiples of hedge heights (six foot hedge). Note (bottom curve) that there is a decrease in percent of crop yield close to the hedge due to shading, but that as wind speed and evaporation decrease there is a large area of increased yield. The large area increase is greater than the shaded area decrease. (from Hedges by Pollard et. al.)

The demise of an ancient whitethorn hedge. The stone-sod hillock can be seen below the fallen trees.

Though the stoat is nocturnal and hunts at night, it is quite commonly seen during the day. It is unafraid of man and I have drawn one out from a hedgerow by squeaking like a wounded mouse. Stoats can climb hedgerow trees with ease and no doubt prey on nesting birds; however, their main prey are mice, rats, and rabbits.

Dogs, cats, and barn owls seem to be the stoat's primary enemy, which is why the little whithred is commonly found wherever piles of rock or hedgerows provide cover from its enemies. However, the stoat is well able to defend itself. One story from Burtonport, County Donegal, tells of a pack of stoats that killed a collie dog and chased away its terrified mistress. Nevertheless, the Irish stoat is the farmer's best friend, as it keeps the rat and mice population in adjacent fields under control.

I believe that the hedgerow enclosure of the lands by independent farmers and the subsequent buildup of the stoat population all over England and Ireland helped bring the black plague under control. The disease, spread by rat fleas, disappeared from England during the centuries of enclosure.

Besides the contribution of the hedgerow to insect and rodent control, its practicality for slowing down strong winds, and thus wind-born erosion, is well documented by research in England. A wire fence offers no shelter at all, and a solid stone wall creates a zone of turbulent wind to the lea. The hedgerow filters the air and turns a strong wind into a gentle breeze with little or no turbulence (figure 5, page 102). urprisingly enough, such a permeable barrier provides more shelter in proportion to height than does a block of high woodland. Like a stone fence, a woodland is a solid barrier and generates turbulent gusts in adjoining fields. A main effect of lower wind speed, besides wind erosion control, is a decrease of evaporation and thus more precious water preserved for the crops.

I have no figures for Ireland but in the book *Hedges,* authors Pollard, Hooper, and Moore (see annotated bibliography) calculate that pre-World War II England had 620,000 miles of hedges, and that between 1946 and 1963, five thousand miles per annum were grubbed out in order to increase acreage (for quick profits): a total of 85,000 miles or about one-seventh of all of England's hedgerows in 17 years. Was it worth it for the slight gain in acreage? Lately, Irish and English newspapers report complaints of rodent problems. Wind-blown erosion has become commonplace in East Anglia and Norfolk where vegetables are grown on the peat fens—called "blows."

I hiked back down the cliffside path. At the base I came to a cut through a hedgerow mound where a new gate was being installed. I recalled a drawing in that classic of agricultural folk writing *Irish Heritage* by E. Estyn Evans (figure 4, page 93), of a mound hedgerow cut through in cross section just like the one that stood near by the trail. I bent down and examined the internal anatomy of the sliced hedgerow. Sure enough—the roots of the whitethorn turned inward away from the ditch and grew through the cracks between heaped rocks. Why? Was there a strange and magic force in those mound-covered fieldstones, I asked myself as I hiked back to Belleek between the walls of my Irish hedgerow zoo?

7

THE MAGIC BIRD

THE ROYAL HOTEL on the Main Street of Boyle is a good place to stop for a night's rest. It is one of the pleasant little West of Ireland hotels that has not been prostituted by over-commercialism.

From the dining room window of the hotel one can watch the tiny, restless grey and white dipper bobbing up and down on the river rocks. This waterproof bird dips low before plunging into the swift current of the river to feed on the bottom. Though unlike sea birds and ducks, the dipper does not have webbed feet, it is at home in and under water. The Irish dipper is distinct from the British bird with its darker upper parts and less chestnut coloring. It is a true native creature.

As I watched from the window I wondered what magic is in its feathered body that allows it to run along the river bottom through the raging torrent of the Boyle. Of course, as Ireland is the land of magic, why should it not have a magic bird? There are magic glens and castles, mysterious round towers, and an-

The little Irish dipper defies the swift water of the Boyle River.

Carrowkeel mountains near Boyle. The Carrowkeel limestone ridges have much in common with the Burran. It was here that we searched for the megalithic village in the fog.

cient megalithic stone structures scattered all over the countryside of Ireland. That is why my wife and I were stopping at Boyle. We had come to see if we could locate the ancient megalithic town and cemetery on the summit of Carrowheel mountain in the Bricklieve range near the village.

The road from Boyle to Sligo crosses the Curlew Mountains and skirts Lough Arrow between the west bank and the cliff-bound Bricklieve Mountains. Carrowheel rises from the lake shore directly behind a roadside pub appropriately named The Mayflie Inn.

Robert Lloyd Praeger, the famous Irish naturalist, discovered the megalithic village on the flattopped limestone ridges of Carrowheel Mountain. He describes the ancient site in his classic book on the Irish countryside (see bibliography).

To reach the crest of the limestone hills you must drive west through the pretty valley of Beal an Atha Fada which separates the Curlew range from the Bricklieve range. Carrowheel is truly a mystic spot and one of the strangest high plateaus in all of Ireland. On that grey October morning, as we turned our little Ford up the steep dirt tracks, a dense white fog rolled up the valley obscuring the precipitous rocky sides above our heads.

The summit of the plateau, somewhat like our own western mesas, except that it is covered with heather, is peculiarly cut and lined with steep escarpment-sided clefts that resemble giant ditches running from northeast to northwest. The dirt tracks followed the bottom of one of these clefts, which grew increasingly shallow near the top of the mesa.

The cliffs were partly obscured by the fog but once in a while a rocky ledge protruded from the white vapor, giving the appearance of a huge stone lintel floating on a cloud of mist. As we approached the summit I could hear a little kestrel calling from some invisible lookout far above. The kestrel is one of the few living creatures that hovers in the air suspended by its forward thrusting wings and down-thrust fan-shaped tail.

The symbolism of the stone floating in the mist and the calling kestrel did not escape us. Almost everything I had ever read

A small Stonehenge in Ireland. One of the best preserved and unique stone circles in Ireland is Drombeg about 1.5 miles east of the little village of Glandore, County Cork. How were the circles utilized by the ancients? One thing I have discovered for certain is that it involved the science of sundialing.

about ancient stone structures contained speculations as to whether or not the ancients understood the principles of levitation. Scientific, mystic, and occult literature is filled with such musings but none gives the remotest hint of how such "magic" might be accomplished. [See also my book *Ancient Mysteries, Modern Visions—The Magnetic Life of Agriculture*.]

The most famous of all stone monuments is Stonehenge in Wiltshire, England, which dates from before 1600 B.C. Those stones are fixed in place and so huge that the principles of kestrel aerodynamics could not possibly apply to such heavy stones—or could they? At that time I put such thought of levitation from my mind since I was more interested in the legends and history of Ireland than the magic of the ancients.

My wife and I never did reach the stone city discovered by Robert Lloyd Praeger. The ghostly mists of Carrowheel were too dense for even foot navigation. We were afraid we might fall off some rocky escarpment if we walked. By the time the sun burned away the fog, it was time to leave; our relatives were expecting us in Belleek.

Since that day on Carrowheel so many years ago, my wife and I, utilizing Sean O'Riordain's little handbook, *Antiquities of the Irish Countryside*, have visited many similar megalithic sites, dolmans, crannogs, and standing stones scattered around Ireland. Ireland is indeed the land of stones and sky!

I have never yet visited one of those ancient stone monuments that it did not excite my mind to wonderment and even expectation. Expectations of what? Better days ahead? New insights concerning creation? A brighter life? I have not found out for certain but for me those stone-loving people emerged from the darkness as fascinating vibrant beings.

Historians call those beginnings the Stone Age. Moderns use the term condescendingly to emphasize that our high energy, inorganic metal age is superior to their low energy, more organic stone age.

Most scholars call the period between the megalithic age of the stones and our present-day age of metals the little understood Celtic period of expansion in Europe and western Asia.

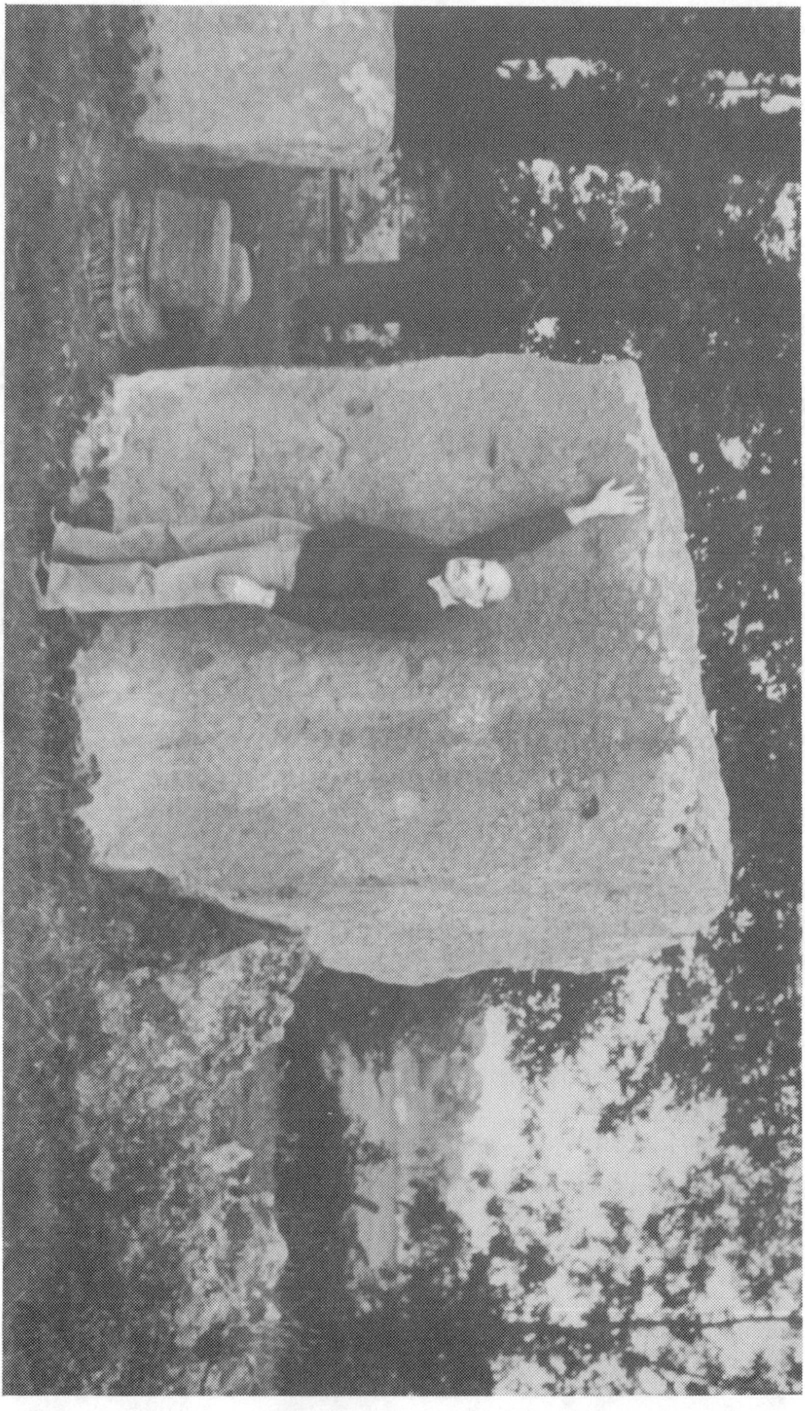

The largest stone on the Lough Gur stone circle.

Ogham stone with hole at Kilmalkedar church, Dingle Peninsula, County Kerry. The beautiful 12th century church has a Romanesque door, rare in Ireland. Besides the holed stone with Ogham writing there is a very fine sundial in the church yard. Ogham writing is a series of dots and dashes etc. This stone reads ANM MAILE INBIR MACI BROCANN (Mael Inbir son of Brocann).

Ireland represents the last surviving, though fast disappearing, body and soul of that volatile period. The Celts, like the megalithic people, left no written words, so we must depend on the archaeologists for our knowledge of both the megalithic and Celtic peoples.

There are, of course, pockets of Gaelic speaking people remaining in northern France, Wales, and Scotland, but there is one important fundamental difference between the history of those lands and that of Ireland. Ireland was never overrun, and thus little influenced by the Roman Empire. The fact that Ireland was never part of the Roman sphere of culture is of great significance to the historical development of Ireland.

Although, unlike England, there is not a single Roman ruin in Ireland, what does remain from the ancient megalithic culture is far more interesting.

The ancient stone structures of Ireland are mainly gallery graves, mountain top cairns, dolmans (one large table-like stone supported by three or four leg stones), standing stones, and stone circles (standing stones in a ring), of which Stonehenge in England is the most famous and intricate.

Professor Burl, a student of stone circles, in his comprehensive book on the subject, lists over 900 stone circles in the British Isles. [Professor Burl is noncommittal as to their function.] Many are still preserved in their original shape. There are over 200 sites alone in Ireland. In the public's mind they are most often associated with the mysterious religious rites of the Druids, and thus the Celts, since Druids were Celtic priests. The Celts, however, were a woodland and agricultural people, not a *stone working* society.

Mitchel (see bibliography) has documented the history of the megalithic stone age farmers of Ireland. Anyone visiting the rock-strewn, treeless countryside of Ireland today has a difficult time recognizing that less than 6,000 or 7,000 years ago, at about the time Egyptian civilization was beginning, Ireland was completely clothed in a dense virgin forest where bear, elk, and wolf roamed at will. This was the beginning of a period that Professor Mitchel calls the neolithic first farming era of Ireland.

Baltinglass, the most spectacular megalithic passage complex in Ireland, lies about one mile north east of the County Wicklow town of Baltinglass on a high hill. The three passage tombs are surrounded by two huge stone rings and a massive stone wall. The view from the hill is magnificent. I believe these so called tombs were in reality churches for megalithic worship, meditation, and healing. They are shaped cross-like as are Gothic cathedrals.

Generally speaking, archaeologists divide the megalithic, or stone age into three periods, the latest being the neo (new) lithic (stone) or new stone age which followed the mesolithic, or middle, period.

The neolithic people, like the earlier mesolithic, buried their dead in enclosed stone chambers. Neoliths are generally associated with fairly advanced stone age agriculture. Indeed, because their dead were well-protected in rough stone tombs containing certain of their worldly possessions, like the ancient Egyptians we know so much about the Irish megalithic people.

The neolithic farmers were characterized by their use of polished stone implements; the art of grinding stone, bone and horn with sandstone, pottery making and advanced agricultural methods, which included domestication of animals, the cultivation of grain, growing of fruit trees, linen weaving, and the beginning of village life. If we substitute cotton weaving for linen weaving this same so-called new stone age definition can be applied to the ancient Egyptian civilization. Like the neolithic Irish farmers, early dynasty Egyptians were truly stone age farmers. The simple fact is that we know a lot about both cultures because they buried their dead in well-constructed stone tombs with their worldly goods to accompany them in the afterlife.

Undoubtedly future archaeologists will have a very difficult time figuring out our own civilization. Words on stone, Egyptian hieroglyphic and Irish Ogham symbols, last far longer than those preserved on paper. Film and video tape is much less permanent than cotton cloth. The Fourier Transform Computer memory which I utilized with my infrared spectrophotometer gets wiped out at least every couple of months by surges of current from the power company. Every time there is a hot or cold spell, which draws considerable current on the lines, computer memory must be reprogrammed. Computer amnesia I call it. There is far *less* permanence in our modern information storage systems than ever before in the history of mankind!

One fundamental difference between the Egyptian and Irish stone cultures is in the utilization of wood. The Celts, who followed the neolithic farmers in Ireland, were far more dependent on wood than on stone. The neolithic people also utilized wood to a large extent in the construction of their villages, as did the later Celts.

Mitchel makes a very strong point when he states: "It is curious that scholars, because they carried in their thinking the tidy fields of the European plowman and the felling of trees by ax, have so often thought that forests repel agriculture and that open land invited it."

He points out that America itself is the best proof that such a thing is inverse to the truth. American farmers never invaded the thick sodded prairie until they built plows able to rip through the deep layer of thick matted grasses. In the last century forests were not even cleared with ax, other than for house timber, but rather by girdling trees and leaving them to die. Crops were planted between the dead trees. It is far easier to clear and burn forest floor litter that lies under virgin woodlands than it is to cut and grub deep grass or brush. It was not until the invention of the metal plow, capable of ripping through the matted roots of sod, that the prairies and steppes of the world were opened to agriculture. Even the early Celtic metal-tipped plowshare could not accomplish that feat.

Today, university scholars are deeply involved in a study that they label sociobiology. A rather simplistic definition of sociobiology is "the study of the theory that a biological organism, man included, can inherit through genes certain cultural patterns." Nobody, of course, has proved this but it is a fascinating concept. I would like to suggest that perhaps we need to begin another area of study called sociophysics, by which I mean the study of the effect of the physical forces, especially weak physical forces, on the cultural patterns of mankind.

Just as the little Irish dipper uses some weak force of "physical magic" to move its light, delicately feathered body along the rushing torrent of the River Boyle, so also those stone age megalithic peoples must have utilized some of the weak forces of nature to their own advantage. Let us see if we can decode just exactly how those weak forces operate.

As a start we must understand that modern scientists are human beings and thus equally likely to protect their little enclaves of special knowledge with mystic (technical) gobbledygook just as the megalithic intellects did 6,000 years ago. They are also just as likely to have a double standard—one that applies to their thinking, and another for everybody else.

This is not a book for physicists or other scientists, but for people with common sense, scientists included—but only if they are not intellectually confined to their own technical jargon.

As an example of the double standard that applies to the language of scientists and the language of mystics, let us examine the word materialization—a favorite of mystics.

In India the mystic Sai Baba is reported able to materialize sacred dust from his fingers. Several of my scientific friends, Ph.D.s and M.D.s, have visited him and could detect no trickery. Some scientists in the U.S. hire stage magicians, sleight-of-hand experts, to put down such mystics. [The Committee for the Scientific Investigation of Claims of the Paranormal.] Some mystics are, of course, charlatans and are easily exposed. Others like Sai Baba have never been proved to be a practicing sleight-of-hand stage magic. What is happening?

If we go to the *Penguin Dictionary of Physics*, an excellent dictionary of physical definitions, we find the word *materialization* (a word scorned by most conventional scientists) on page 236. It reads:

> Materialization. The direct conversion of energy E into mass according to Einstein's equasion $E = mc^2$ (c= velocity of light), as in pair formation.

If the physicists can mathematically believe in materialization, then a mystic type has as much right to believe his eyes especially since it makes mathematical sense. What is good for the goose is good for the gander.

Let us examine this definition more closely. In order to understand Irish magic my reader must first follow my common sense reasoning. It is, therefore, not my intention to insult my readers' intelligence when I point out the simple mathematical elegance of Einstein's formula, which was deduced (not proven) from physical experiments, e.g. Michelson supposed measurement of the speed of light.

In words instead of mathematics what Einstein said is that energy equals mass (weight is a measure of mass) times the speed of light squared, or mathematically:

$$4 = 2 \times 2, \text{ or, } (E = m \times c^2)$$

If this above formula is true then likewise, to convert:

$$4/2 = 2, \text{ or, } (E/c^2 = m)$$

in other words energy divided by the speed of light squared equals mass. The speed of light, based on Michelson one experiment is 186,000 miles per second. To square it we multiply 186,000 by itself and that is a pretty big (fast) number!

When we convert Einstein's formula using logical mathematics we have placed energy divided by the speed of light squared on one side of an equal sign and mass on the opposite side. We now have a unique dilemma, which is of course demonstrated by the mass (weight) of this book. There would, of course, by the very nature of this formula not be any book at all were not light (wavelengths) traveling at speeds equal to or slower than the *speed of light squared*. In other words the formula *demands* light having speeds equal to or lower than the energy (E, left side). Inherent in this book then, according to Einstein, is the simple fact that as the speed of light squared decreases, e.g. $4/1.5 = 2.6666$ etc., (everything is approximate) mass increases. This of course means that while the Great Pyramid at Giza is quite slow (where mass is concerned), my little few ounce mystic water dipper is a ball of fire (where mass is concerned). This book lies somewhere in between the water dipper and the massive pyramid at Giza.

Of course the physicist will argue that Einstein was talking about very tiny sub atomic particles and not on a molecular (pyramid and water dipper) level.

I do not believe in statues for scientists. Unfortunately there is one to Einstein in Washington, but does that make him a god? Einstein, who was a gentle person, would probably not approve of the statue were he alive, and he would be the first to agree that if $E = mc^2$ is true on a micro-atomic level, it might also work its way up to a macro-atomic (molecular) level.

Let us now look at another word, one that chemists use constantly—*synergism*.

The definition for synergistic chemical reactions, according to *Webster's Third New International Dictionary* is: "Cooperative action of discrete agencies (as drugs or muscles) such as the *total* (my italics) is greater than the *sum* (my italics) of the two or more effects taken independently."

Wow, just a minute professor, you are saying mathematically that $2 + 2 = 5$! Exactly! Chemists, however, seem quite self satisfied with that definition as do we biologists. If a chemical reaction produces energy greater than the sum of its two parts, then the extra energy must be coming from somewhere. Where?

There is of course only one answer and it must of necessity trace directly, or indirectly, back to that source of all energy, the cosmos and that cosmic generator the sun.

All physicists agree that matter and energy are never destroyed; they are only converted back and forth from one to the other (energy to mass [form] and vice versa). A brick of Irish turf is converted by burning (oxidation) into gas and soot, but the sum of gas and soot always equals the mass (weight) and energy of the brick of original turf.

If we believe in classical physics, then we must understand that a synergistic molecular chemical reaction must involve the ability of reacting molecules to operate like little antennae and collect the extra energy from the cosmos (sun). "Antenna" is, of course, a modern and very elegant magic technical word for an energy collector. It is a magic technical word because although we know that an antenna collects energy, and that its efficiency depends on its form—that is, on the shape of its mass, we do not really know how an antenna works.

Because physicists want to protect their status as the high priests of technology, they have made up another magic word, *resonance*, to explain how antennae work.

Simply, if the molecules of two or more chemicals are mixed in the correct proportions and ratios (called alchemy by the ancients), then at any one instant in time and space the correct resonant configuration might occur, whereby the mass has the proper shape (form) to act as an antenna and collect from the cosmos vast amounts of that extra energy. We then have what chemists call a synergistic reaction where: 2 × 2 equal 4 + 1 (extra collected energy). The reaction usually occurs in microseconds and the collected energy is instantly amplified by resonance and thrown out from the molecular antenna, making it seem that 2 + 2 equal 5! This is a physical form of molecular antenna magic.

We may now understand that by shape an antenna is a collector and also an amplifier; that is it can collect and increase energy output by means of that inexplicable mechanism that we call resonance. Of course, if certain shapes can amplify energy, then certain shapes might also amplify (speed up) time, thus increasing the velocity of light to 4/2 so that energy converts to mass, causing materialization. (There is new evidence for such a time reversal)[1]

It is interesting to note that Webster's definition of a synergistic action uses drugs and muscles as an example. Drugs and muscles work at room temperature and are excellent examples of 2 + 2 equal 5! Sai Baba no doubt has muscles in his fingers and may well know of some ancient herbal brew that makes 2 + 2 equal 5 due to the shape of the organic molecules residing

1   A particle faster than the speed of light is called a tachyon. A message sent by such a particle would arrive before it was sent. In *Speculations in Science and Technology*, volume 9 (1), pp 51-59 (1986). I published the first experimental proof of tachyon pariticles (photons).

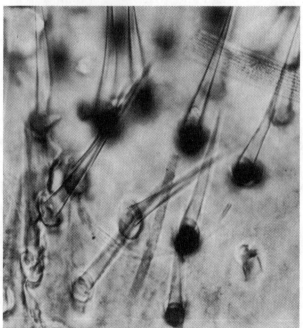

Cecropia moth antenna (top). Small spines (sensilla) occur on the branches, sharp pointed ones are at the base (bottom).

in his finger tips. I, of course, have obtained my information that Sai Baba does materialize sacred dust from several highly reputable scientists and M.D.s, but whether or not he does or does not there is nothing in the process of materialization that contradicts classical physics, or for a matter of fact antenna engineering.

Everything in nature that has actual form can be treated as an antenna, that is, as a resonant shape for collecting some type of energy. The physics applies whether it involves the shape of the great bowl-like form of Ireland (Chapter 3), or the minute spines on plants (trichomes) and insects (sensilla) that occur in nature. It is also true for trees, bushes, and the huge standing stones and stone circles that are spread across the landscape of Ireland, and, of course, for the strange three-legged dolmans that dot the high moorlands of Ireland. It especially applies to tiny molecules.

As I sat by the window of the pleasant River Side Hotel, eating dinner with my wife and watching the magic dipper plunge into the swiftly flowing river, memories of past days on the banks of the winding Erne flooded back.

Long before that beautiful river was dammed up to form a huge lake for electric power, I would walk out on the eel weirs and watch a pair of dippers that patrolled that part of the river. The eel weirs were long, man-made rock projections that jutted out at an angle into the river to funnel migrating eels into traps.

On a quiet summer evening the weir rocks were a good place to sit and watch the life of the roaring river. There were leaping salmon, silver damsel flies on wet shinning rocks, and dippers plunging in and out among the river washed rocks.

I once learned firsthand about the powerful force of such swiftly flowing rivers.

At its source in northern New Mexico the Rio Grande is a white water force beyond the reckoning of those foolish enough to challenge it. My youngest daughter, Colette, son Kevin, and their friends and I were rafting down the turbulent Rio Grande when the rubber raft hit a rock and I was thrown out the back, camera and all.

The mighty river sucked me under so I decided to stay beneath the water, hoping that the force would direct my unprotected body around the rocks. It was useless, of course, for a one hundred sixty-five pound man to fight the current, so I held my breath and was swept along. Sure enough, my head never bashed against a single rock and I was spit up one hundred yards down river in a clear spot, where I was pulled aboard the raft. My body beneath the water had kept up with the raft on the surface. Whenever I think of the forces involved in my rather adventurous river run, I think also of my little Irish dippers.

I assume that the shape of the bird, and perhaps some water flow hydrodynamics principle, allows the tiny five or six ounce bird to run against the river force.

Olive Thorne Miller, in *Birds of America*, says that our Rocky Mountain species (called a water ouzel) can actually fly under water. She calls the bird a "curious fellow" who lives on insects underwater. Its feathers are soft and so thick that its body never gets wet.

It is, to my way of thinking, no more astonishing that materialization should occur, or that some certain molecule or stone structure should act as an energy collector (antenna), than that God should evolve a little creature that can defeat the mighty force of the River Erne. In nature there will always be inexplicable phenomena and magic creatures like the little dipper at Royal Hotel.

8

THE WITCH FROM COUNTY CLARE

WHEN I AWOKE that October morning at my daughter's house on top of a hill near Ennis, I opened the drapes over the huge sitting room window. The oblong window gave a postcard view of the green rolling hills of County Clare spread to the east.

The wind was blustery and it danced along the hedgerow that edged the yard. One cannot, of course, see the wind dance, only the effects of the dance as the hedgerow limbs twist and bow before the invisible force. The County Clare wind is magic—the hedgerow was proof enough of that. The un-trimmed tops of the whitethorn undulated in waves like the legs of dancing chorus girls, while the smaller blackberry bushes bowed and twisted under the weight of their ripe fruit.

Blackberries are one of two good reasons for visiting Ireland in October despite the wind. It is the time that we can take our grandchildren blackberry picking. The second reason is rain, or rather the lack of rain. There is an old Irish folk saying, printed on many souvenir calendars, that proclaims; "There are twenty-one fine days in October." So far we had had eighteen fine days and, since the skies were a marble blue, we thought that perhaps we should not tempt fate with only three sunny days still due us that month. Instead of blackberry picking with the grandchildren, my wife and I would drive to Feakle and try to find the cottage of Biddy Early, the famous Irish healer. Then, if we had time we would circle west and drive across the great limestone plateau of Clare called the Burran.

Biddy Early, an enigmatic country healer, is known to folklore and song as the "Witch of County Clare." Healers were often labeled witches. It was a designation which in past centuries could lead to death at the stake. In the Middle Ages, and even into the eighteenth century, there was a time for healing and a time for burning.

Healers were secretive, and women healers especially were looked on with considerable suspicion. Fear of magic and jealousy often led to accusations of witchcraft. Strangely enough, though many thousands of so-called witches were executed in Europe, I have never found even a single documented case of an Irish healer being condemned for witchcraft, despite the fact that Irish witches were as secretive and feared as their English and Continental counterparts.

It is difficult to decide if such healers were secretive because they were suspected of witchcraft, or were suspected of witchcraft because they were secretive. Most modern treatises on the subject assume that they were secretive because they wanted to protect their professional status, a non sequitor better applied to modern professors than to medieval witches who never accepted money for their knowledge of healing.

Biddy Early could certainly not have survived on gifts of produce had she not been supported by the three men she married. After her first and second farmer husbands died, she married a third time to a handyman-laborer.

Biddy Early was born in the year 1798 in the townland of

Faha near Feakle. Despite the tragic year of her birth she survived to grow into a beautiful red-headed teenager with a pleasant personality.

Ninety-eight was the year of the hopeless revolution led by Wolf Tone, the great Protestant Irish patriot. Well over 30,000 men, women and children, armed only with pikes and pitchforks, fell before the troops of the British Crown.

Although Biddy was christened Bridget Ellen Connors, she took her mother's name, Early. In Ireland the gift of healing was considered to be handed down from father to son or mother to daughter. In the latter case the maiden name was more often than not retained. Biddy's mother, Bridget Ellen, was known to have a keen knowledge of herbal cures and probably instructed her daughter.

Biddy's mother died when she was sixteen. Her parents, as was usual, did not own their little plot of land. Most of the daughter's and father's labors went to pay the rent. A couple of years later her father died of "the fever," probably typhoid, and Biddy was thrown off the farm.

In those days the population of Ireland was around eight million, over twice the modern census figure. Biddy, like most dispossessed widows and children, was forced to take to the highways. Landless males were usually able to hire out as farm laborers for their keep, but women were forced to become nomads. Ireland in winter is a cold and damp environment and few survived for very long.

Since no small farm could afford an extra woman, they moved from household to household helping out with daily chores of milking, carting the turf, cooking, etc. Such women were taken in out of compassion alone, but apparently Biddy's gentle disposition did not allow her to impose on her neighbors for any length of time.

She eventually obtained a servant's position with a landlord named Sheehy. He turned out to be a tyrant and made life intolerable for his help. When he was murdered by some of his tenants, who also hated him, Biddy again took to the roads. For a while she lived in a miserable workhouse, but eventually

The inside of Biddy Early's cottage—the lonesome plate on the Irish sideboard tells the tragic end of this once preserved cottage—a stark reminder of the thoughtlessness of modern technological man.

she met and married a widowed farmer, Pat Malloy, who had a son by his first wife. Despite their age difference she was extremely happy for a few years on his farm at Gurteenreagh.

When her husband died, she married his son John who had gotten on well with his stepmother and considered her a good friend. This, of course, allowed her to remain on the farm at Gurteenreagh. Under harsh rural conditions marriage was a matter of survival and seldom if ever the result of romantic love. John died of alcoholism in 1840. Apparently Biddy's fame as a healer and herbalist was spreading far and wide and John usually spent his evenings with friends around the fireside overindulging in the illegally brewed poteen (potato whiskey) that grateful patients left with Biddy.

During this second marriage her only child, Pat, by her first husband, left home for seven years, after which he returned to live with Biddy and, according to folklore, brought her the mysterious blue bottle that she used in her healings. It was said that the fairies gave it to him.

After John died she married a laborer named Tom Flannery, who built the little two room cottage we were seeking in an area called Dromore near Feakle.

Biddy lived to a ripe old age in that cottage on Dromore Hill in the township of Kilbarron. Dromore Hill overlooks a small pond known today as Early's lake. This so-called lake is in reality only a low marshy body of water that collects the drainage from the surrounding hills. On her deathbed Biddy asked her son to throw her magic blue bottle into the lake.

Little or nothing is known of Biddy's cures—only that they worked and that she had certain magic powers. Some believed she had the ability to forecast the future of those who came to see her. One such visitor was the great Irish patriot, Daniel O'Connell, who sought her advice on election matters.

It is known that she utilized standard herbal plants such as thyme, plantain, nettle, and moss in her curative mixtures. All of those plants have medicinal properties. Biddy was seen picking green tufts of moss from the water wheel at the mill at Ballylee beside the towerhouse that William Butler Yeats later

The small pond below Buddy's cottage where her son threw the mysterious blue bottle.

converted into his own home (see Chapter 2). Ballylee is near Faha where Biddy was born. I always considered it strange that Yeats, Ireland's most mystic poet, never wrote a word about Biddy, who was Ireland's most mystic healer. He must have known about her.

Country cures known to rural folk were usually handed down by word of mouth, thus most modern treatises usually leave out some necessary component of the concoction that makes it work. For instance, Meda Ryan, who wrote Biddy's biography, says that Biddy's cure for a sore throat and also for what country people call "tightness of the chest," was made with "little trouble" from the common turnip.

As far as it goes that is true. Our good friend Teresa Pollard, formerly from Listowel, County Kerry, now living in Florida, described in detail the turnip cure used on her as a child. The turnip is cut in half and the middle portion is half scooped out. The hollow is filled with fresh bee honey and the "turnip bowl" of honey is left for several days to mix with the juices of the turnip. The result is a very watery syrup to be taken a few times daily. I know it works, for I once tried it myself. I must, however, caution the reader who is trying such a country cure, for like all cures it does not work for every chest problem. Besides, I do not want to get the reputation of being a male witch.

Modern man no longer burns witches, instead, we burn their books. What is sad about this rigid mentality is that instead of getting a reputable chemist to find out what accomplishes the cure, the technique is summarily condemned as superstition.

My generation may well be the last to have direct contact with reliable people who have been healed by such remedies and remembers the essentials of the cure. Apparently modern man does not have the poetic soul necessary to pass such information along with exactitude—tape recorders notwithstanding. The donkey cure is one good example of exactly what I mean.

This cure for mumps is, on the surface, one of the most ludicrous cures ever described to me. I have only found one Irish book that mentions it, no doubt because even the most holistic minded folklorist would be hard pressed to maintain that it works.

Credibility is important to a scientist. Nevertheless, I shall risk it because the donkey cure for mumps is entirely consistent with good solid-state physics, and I know at least two County Donegal people who maintain that their pain subsided following the treatment. Mumps, which is seldom fatal, but is very painful, is a glandular swelling in each cheek in front of and below the ear. According to Dr. Patrick Logan, author of *Irish Country Cures*, the winkers (halter) of an ass is tied around the sufferer's swollen neck. The patient is then led by the halter to a stream and given a drink from the flowing water. It is important that no vessel be used to dip up the water, although the human hand is allowed. The stream must run north and south. My sister-in-law, who received this treatment from the local healer, James Quinn, informed me that he scooped up the water with hands which were, to put it simply, stained with the earthy residue of his agrarian trade.

One essential part of the cure left out of Dr. Logan's description seems, at least to me, more farfetched than the halter or north-and south-running water. This vital procedure, included in verbal descriptions given to me, was invariably emphasized by those who had taken the cure. It should be difficult to forget, for the patient is required to crawl under the belly of the ass!

Now conjure up in the magic portions of your brain called imagination the sight of a young lad or lassie with a rope halter around his or her neck crawling back and forth under a donkey standing near a moorland stream. How utterly ridiculous—just the sort of thing an ignorant peasant would resort to! The trouble with this sort of modern contempt is that it assumes that the brain of the ancient, from whom these cures were handed down, was less developed than ours and that he never experimented with natural processes. Yet, four or five thousand years in human evolution is as nothing. The simple fact is that the ancient human mind was little different from the human mind of the twentieth century. The ancients practiced low

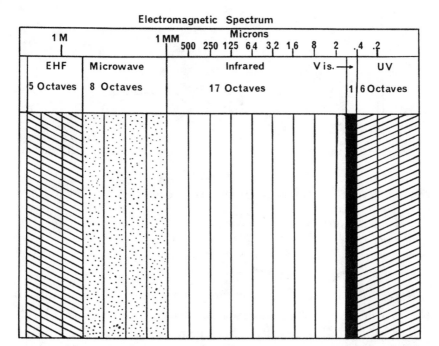

Electromagnetic Spectrum

1 M		1 MM	Microns											
			500	250	125	64	32	16	8	2	.4	.2		
EHF	Microwave		Infrared							Vis.→			UV	
5 Octaves	8 Octaves		17 Octaves									1	6 Octaves	

Figure 6. The electromagnetic spectrum from extremely high frequency (EHF) radio (meter long waves) to UV (sub micrometer long waves). The infrared between visible (0.4 to 0.75 micrometers = microns) and microwave (millimeter long waves) is the largest part of the electromagnetic spectrum and covers 17 octaves of radiation.

energy agriculture and medicine, we practice 110V high energy agriculture and medicine. Which is best? Neither and both! What is needed is a marriage of what works best with our modern high energy systems, and what works best with the ancient low energy systems. The mule pulling a plow is definitely less efficient than a regular sized John Deere tractor, but it is most certainly better than a fifty-ton monster that compacts the soil and rips up thousands of acres for a fast profit. Modern man has come to believe that bigger is better and that everything has to be plugged into a 110V socket; unfortunately for modern man, nature does not work that way.

Let us examine the donkey cure in light of what we know about solid state physics and modern radiation theory. Most of what takes place in nature occurs at room temperature in the

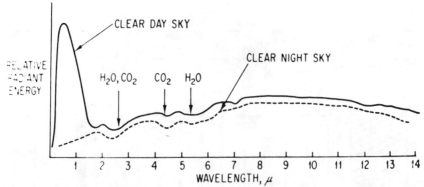

Figure 7. Visible and infrared radiation from the day and night sky. H_2O and CO_2 absorb a certain amount of the far infrared. Night sky radiation peaks at 10 μm and takes the form of a long low curve. The day sky peaks in the visible region (tall curve) with a long infrared tail similar to the night sky. The long infrared curve of night and day is called blackbody radiation since it is not visible as is the sun (hot) generated visible peak.

infrared portion of the electromagnetic spectrum and also in the visible and UV region (see spectrums, figures 6 & 7). The UV, visible, and near infrared are generated by the sun millions of miles away. It requires a lot of heat to generate such frequencies, which is precisely why you burn your hand if you touch a 60W light bulb, or the earth would fry if it moved any closer to the sun. Not so for the far infrared, between 7 and 100 μm (micrometers) in wavelength. Such frequencies are generated at what we call room or earth temperatures. We walk around bathed in a sea of infrared radiation. This being so, why then do we put so much more money into research on high energy radiation than in the low energy infrared where biology is concerned? Two reasons, I believe:

1. We believe that forcing systems by hitting them over the head with high energy is better than letting nature take its slow but sure low energy infrared course, e.g., kill the cancer with massive doses of poison, or high energy radiation, since nothing else could possibly work. Of course, the obvious side effects of such a cure are that you kill the healthy cells also. It is always the ultimate of hedging when a doctor says there *might* be serious side effects from such a treatment, when he knows full

well that it is unavoidable—dead cells are dead cells!

2. Most modern infrared research is done by the U.S. Department of Defense and involves systems designed to spy on the earth. These systems developed for spying are not available to researchers in agriculture and medicine.

Although the Defense Department scientists monitor my infrared work there is no reciprocal flow of information. My discoveries have been made by me working alone.

Let us see how what I have discovered in nature might explain why the rope halter donkey cure really works for mumps. Let us examine the entire procedure to see if there might not be some kind of natural energy at work, after all.

A true naturalist does not just bird watch, or collect and classify plant or insect life. A real naturalist, or natural philosopher as such a person was called in Darwin's day, tries to explain how nature works; he not only examines what he can see with his senses but also what he cannot see but suspects to exist.

One of the greatest naturalists of all times was the Irishman John Tyndall. He was born in Leighlinbridge, County Carlow, Ireland, the son of a Protestant police sergeant. John's father sent him to what was called a hedge school. Hedge schools developed in Ireland during the days of suppressive English penal laws. Because the laws forbade a Catholic from attending school, the ingenious Irish organized secret schools that were "hidden behind hedges," and taught by Catholic schoolmasters in rural areas. When the English repealed the penal laws in 1802, the hedge schools developed into the private Catholic school system.

The Catholic schoolmaster at Leighlinbridge was a certain Master Conwill. As he was considered the best schoolmaster in County Carlow, Tyndall's father, no bigot, enrolled his son in Conwill's school. His own words to the Protestant Dean of Carlow, who protested the sergeant's decision, were: "Rev. Sir, if Conwill taught on the altar steps, I would send my son to him, for I have no doubt he will receive from Conwill a sound secular education that will fit him for life."

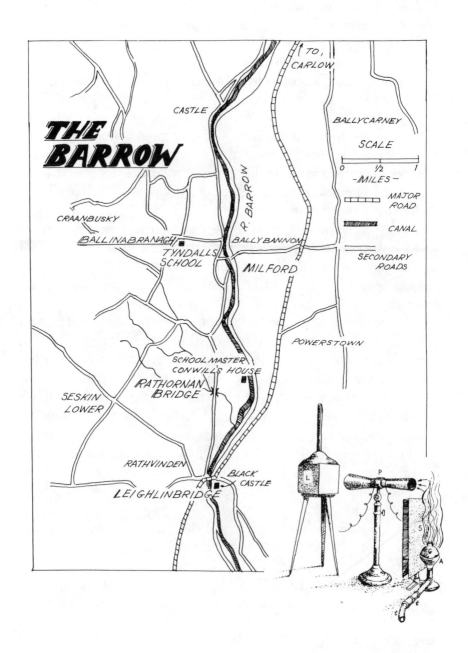

John Tyndall 1820-1893

Sergeant Tyndall's words were prophetical, for after John left Ireland, he taught for a short time in a Quaker school in England, then went to Germany where he obtained a Ph.D. in mathematics under the famed chemist Bunsen. He returned to England and eventually became secretary of the Royal Society.

He was a generalist, as were most early naturalists, so that his discoveries ranged through physics, chemistry, and biology. He described the action of penicillin on bacteria over half a century before it was rediscovered. He also described that the sky is blue because short wavelengths of blue light from the sun are scattered by small atmospheric particles. His foremost contribution, however, is his discovery of far infrared radiation, and

Corn Earworm moth (Heliothis zea) going to infrared coherent lines from scent "pumped" by light in the window at night.

Moth in Irish pub going to the alcohol molecules in front of the singing Irishman's breath. Note the moth does not go to the lights but to air space in front of the Irishman's mouth where the alcohol scent is "doped" with ammonia from his breath. Ammonia is a synergist to plant scents so its infrared lines are strongest there.

The Witch from County Clare 139

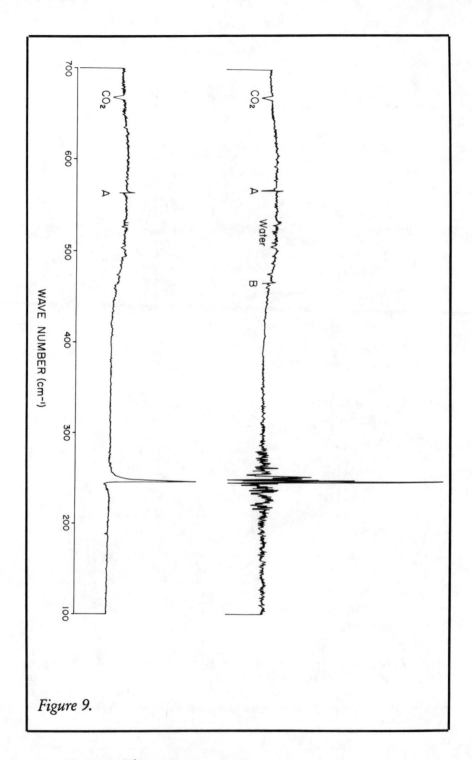

Figure 9.

also the development of a unique instrument to detect it (see map). He wrote of his discovery of far infrared in a book entitled *Heat as a Mode of Motion*. His discoveries cover six volumes, yet despite this he is little honored as a scientist.

Heat as a Mode of Motion seems a strange title for a book by a natural philosopher. What did he mean? In 1800, Sir William Herschel discovered that invisible radiation was emitted by hot objects, such as the sun. It is the temperature, heat or warmth, that causes molecular motion. The new radiation was called infrared because it was below (past) the red in the one micrometer region (1 μm, see spectrum, figure 6, page 133). It was Tyndall, however, who designed a sensitive thermocouple system (called a spectrophotometer) to detect infrared at earth or room temperatures (warm, not hot) called far infrared (10 μm region, see spectrum, figure 6, page 133). He blew various scents, such as rosewood, ethanol and ethaline, across a metal ball with warm water in it and looked at the energy on his meter generated by the flowing gas. This type of infrared is called "blackbody" infrared (see figure 7, page 134) because it is very broadband and all the photons (particles) of infrared light go out randomly. All the little soldier molecules are out of step and scatter in all directions.

Physicists later learned that some radiation, e.g., radio and laser light, can be forced to emit in very narrow bands, with all the little soldier molecules in step (see figure 9, page 140). In other words instead of being broadband and incoherent, all the soliders are in step (called resonance). The beauty of coherent radiation is that it can be collected with an antenna the same length as the wavelength of the light, radio, or infrared wave being emitted. You can tune to it (see figure 10, page 142).

In 1956 I studied the corn earworm moth (*Heliothis zea*) at night and found that the female, which lays her eggs on corn silk, could find the scent around a light bulb even if her eyes were painted out with opaque black paint (see photo, page 138). This left the antenna, with all its hundreds of strange shaped spines (called sensilla), as the reason for moths going to light— *not the eyes*.

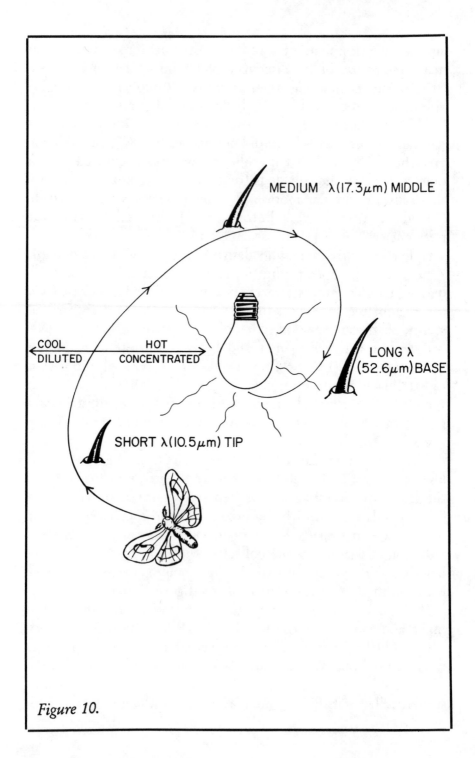

Figure 10.

My discovery contradicts thousands of years of belief going back to Aristotle that insects are attracted to visible light because their eyes "see it." Yes, they can see the light, but are no more attracted by what they "see" than we would be. What happens is exactly what John Tyndall discovered over one hundred years ago, but what has been ignored for over a century.

The hot light bulb, or sun, is putting out more visible, one micro-meter near IR, and far IR (10 μm region) than Tyndall's warm metal ball. That energy, called incoherent blackbody (see figure 7, page 134), goes into whatever scent e.g., corn for the earworm, or cotton for the boll weevil surrounds the light bulb and is stored (absorbed or taken in) by the little scent-oscillating molecule (see figure 11, page 144). The scent is "primed" so to speak, by the large amount of visible and infrared incoherent light emitted by the light bulb. Tyndall realized the little scent molecules in the air took in (absorbed) much visible light and infrared from the environment and then suddenly emitted infrared when they collided with the slightly warmed sides of the metal ball. That is why he called it an emission spectrophotometer.

Over one hundred years later in a paper I wrote in 1956, though not published until 1962 because of small-minded scientists called "anonymous peers" (censors, in other words), I pointed out that under certain conditions of collisions the wavelengths came out in narrow band coherent frequencies. (I call them maserlike—like a molecular radio laser.) [Figure 9, page 140, a maser is a molecular *radio laser*. Certain molecules, e.g. ammonia, can be stimulated to emit coherent radio waves.]

The conditions are such that not only does the scent need to be "taking in" (absorbing) visible, UV, or infrared of the blackbody type (broadband) from the environment but it also had to collide with a special vibrating surface, that is a surface such as an insect vibrating antenna. I patented the first such system (patent no. 3997785, Callahan, Dec. 1976).

To prove this theory conclusively in this patent, and certain scientific papers, I detected far infrared wavelengths that

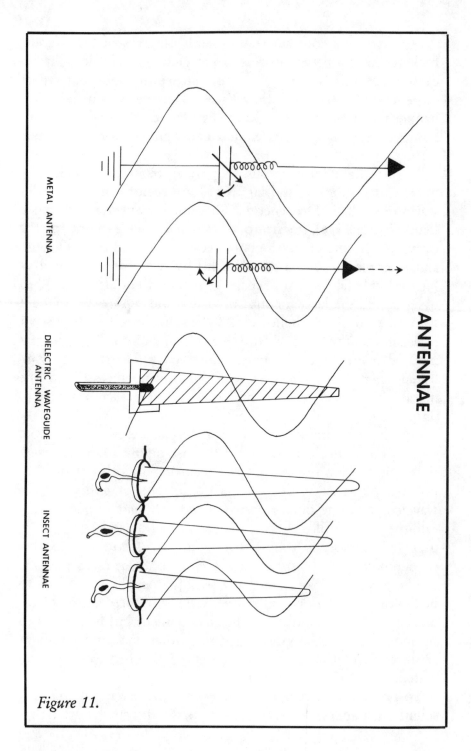

ANTENNAE

METAL ANTENNA

DIELECTRIC WAVEGUIDE ANTENNA

INSECT ANTENNAE

Figure 11.

matched the length of the actual spines (sensilla) on the antenna of the male cabbage looper moth (*Trichplusia ni*). The scent I used was the female sex scent called a pheromone. I did the same thing for moth food, such as ethanol which comes from fermented nectar and honey. I found exactly what the theory predicted when I vibrated a cotton tip applicator at the same frequency that a feeding moth vibrates its antenna, and blew ethanol across the vibrator. I detected maserlike coherent radiation from the ethanol atmospheric mixture flowing across the cotton tip. This is a gaseous maser system, but there can also be liquid flowing maser systems, e.g. A few molecules of a drug or hormone flowing through capillaries and veins and generating *coherent scatter frequencies* in our blood vessels. There is no question at all that just as these insect sex and plant scents control the behavior of my little moths, and other insects, that the same type of coherent infrared radiations, that are stimulated by flowing molecules in collisions with cells lining our own lungs (coherent far infrared), control life processes at room temperature. I have proof of one such important coherent wavelength at 40 μm. This wavelength dimension is within range of the average diameter of human alveoli (20 to 40 μm). The figure demonstrates a frequency generated by colliding the air (atmosphere that all animals breathe) within an artificial reticulated surface of the same dimensions as the reticulated alveoli (cell lining) of human lungs (40 μm wavelength).

Now, my dear reader, you can understand that just as high energy coherent radio must control our satellites millions of miles away, coherent low energy far infrared must control life processes when it is emitted by scatter mechanisms close to a cell wall (or insect antenna). You can likewise understand why the donkey cure may well work and in fact according to two people I interviewed, *did* work.

Before we explain how the donkey cure works we need to cover one more neglected bit of scientific history.

A little known English natural philosopher named Stephen Grey was the first experimenter to discover electricity flowing not in metal, but in the fibers of the hemp plant. Little is

Another strange cure that works. The backache chair cures lower back problems of certain kinds. My daughter Cathy tries out the backache chair in the brush thicket of the Burran.

known of Stephen Grey except that he was a member of a Chartist group (social reformers) and experimented with wax and string. He found that if he rubbed a glass rod with a long cord of hemp tied to it he could move a little piece of gold leaf at the opposite end of the cord. He called his newly discovered moving force electrification. We know that Benjamin Franklin read about Stephen Grey's experiments and this led to the famous string-kite-key lightning experiment. We now understand that weak electric currents flow quite easily along a plant fiber.

Consider then the following. Irish farmers in olden days were totally self sufficient and seldom wasted money on such things as leather halters. Straw or hemp rope was utilized as the donkey halter on most farms.

Thus the sick child was led to water by an excellent plant conductor of low currents placed in a loop around his or her head and neck and over the swelling. Such a hemp or hay halter is doped with the organic salts of the sweat of the donkey, so it is, no doubt, a better conductor than new rope. The rope is also in the form of a loop over the head and neck and, like a loop antenna, will collect radiation along its face (not edgewise).

Next, our healer finds a stream lined up north and south, i.e., magnetically oriented, and gives the suffering patient a drink directly from the stream, from his own earth-tinted hands. The highly conductive water coats the patient's throat so that we have a throat "conducting loop" within a plant "conducting loop" separated by the swollen glands. We now arrive at the most ludicrous part of the cure. The poor country lad or lassie is told to crawl three times beneath the belly of the donkey!

Before we look down our long, skeptical 20th century noses, let us measure the radiation from a donkey's belly. We find that it emits tremendous amounts of far infrared radiation. In fact I have measured, not a donkey's, but a pony's belly emission, and it pegs a radiometer needle right in the middle of the 10 μm far IR spectrum [9.84 μm at a pony's belly temperature of 82°F.]. The belly of a donkey or pony puts out more energy in the far IR than the best X-ray machine proportionately in the X-ray region.

We all know that infrared bulbs are used to heal sore, swollen muscles; unfortunately, these bulbs are called heat bulbs because they are very hot. Far IR has been mislabeled heat radiation. Visible radiation and near IR radiation (1 μm region) are really heat radiations, as it takes a very hot (incinerating) heat to generate them. Far infrared should be called luke warm radiation, as the donkey's belly, or my belly, generates IR at 82°F. Far infrared might even be called "cool radiation." A donkey's belly reflects most of the visible and near infrared (1 μm region) which the sun or a heat bulb gives off. Conversely, it generates (emits) far IR in the 10 μm region. It is a perfect 98% efficient far IR transmitter.

We see now that we have *two* IR wavelengths coming from the donkey's belly—at the far IR emitting at 9.84 μm, and the near IR, peaking at 1 μm, which is reflected from the belly (figure 6, page 133). The 1 μm is, of course, not hot either, because it has traveled all the way from the sun, so has cooled down. We have the radiation without heat. In other words, the donkey's belly is as reflective of 1 μm radiation in the IR region as a silver mirror would be in the visible region. Though if I touch my 60-watt desk lamp I burn my fingers, I can reflect the visible energy very efficiently with a mirror and not burn my eyes or body.

We now understand that we have an organic machine, the donkey, directing its 1 μm sun reflected IR and its 9.84 μm generated far IR at a plant loop antenna made out of thousands of natural fibers (fiberoptics) called waveguides. Infrared waveguides are always made out of dielectrics (such as plastic or plant materials) rather than metal. The radiation spins around the plant conducting loop antenna which amplifies it, as does any properly designed antenna, focuses the IR across the throat through the tissue of the swollen glands, and spins away down the highly conductive throat that has been doped with weak electrolytes from a dirty hand and a flowing stream. Seen in terms of modern solid state *low* energy physics and modern waveguide (dielectric) antenna engineering, this donkey cure makes more sense than using poisons, e.g., chemotherapy to treat cancer.

As I thought deeply on this unique cure, I soon came to realize that just as an insect must shake (vibrate) its antenna to modulate (make detectable) the frequencies from a plant or sex scent, so some fevers will heat up the body and slightly shift the coherent far IR frequencies causing shivering, which modulates (shakes) the far IR energies as they shift with temperature.

Thus the environment, in which the "bad particles" (viruses) live, changes drastically so they can no longer survive. [I used this principle to write a theoretical paper (with a working model of the virus) on a cure for AIDS—by disrupting the virus in the same manner (soft X-ray-UV region) as the donkey cure.] Good doctors are finally coming to understand that the shivering that accompanies a fever is a curative process, and although the doctor wants to keep the fever from going so high that it kills the patient, he does not want to completely suppress it.

My wife and I finally found the house on Dromore Hill. Two years ago my son-in-law Michael and daughter Cathy had taken photographs of it for me and the thatch, although very old, was still intact. The house had obviously been well cared for by the last generation, as Biddy's furniture was still in place. But today the thatch has been neglected and has fallen through into the room, and the old furniture is exposed to the elements. Like Biddy's blue bottle, in which she no doubt mixed her diluted herbal cures, the little thatched-roof cottage is being "thrown away" by the modern Irish society.

In a little valley on the limestone plateau called the Burren, is the thatched-roof cottage where Michael Cusack, founder of the Irish Football League, was born in 1847—about the same time that Biddy Early was moving into her wee cottage on Dromore Hill. Michael Cusack's cottage is being carefully restored, but Biddy's cottage is in ruins despite the fact that Feakle is a very prosperous farming and sport fishing center.

As I walked back down Dromore Hill, after an hour or so

photographing and drawing the floor plan of the cottage, which strangely enough sits in a fairy ring of whitethorn trees, I wondered at modern values that glorify a sport, albeit an excellent sport, yet totally ignore the remarkable souls of this world such as the great Irish scientific "magician" John Tyndall, and the great Irish folk cure "magician" Biddy Early. Tyndall's genius was that his low energy research explains, and makes rational, Biddy's low energy medicine.

When I reached the end of the path down Dromore Hill I turned and gazed once again upon the simple cottage. Biddy, who never took a penny for her loving cures, suffered greatly in this life and was cursed by those who were jealous of her God-given talent. What we know of her has been preserved, not by the Irish Tourist Board, but rather by the delightful folk tales of the simple rural farmers who loved her.

When my wife and I finally drove away to the Burren, there were tears streaming down my cheeks for the witch of County Clare.

9

SMALL WINDOWS, HIGH DOORS

A LOW MELANCHOLY SORT of thumping echoed across the still lough. White mist rose from the surface of the water. The thumping did not come from nearby. The waves were far too small to smack the boat, and we were moving too slowly for the dripping oars to smack the waves. The ruddy faced Erne fishermen were in no hurry to leave me on Devenish Island alone before the mist would clear.

It was a strange sort of whooshing sound, almost like the noise from the puffing hole at Poll na Brioscarnach on the south coastal cliffs of Aran Isle. The sea water at the puffing hole sucks in and out and puffs its whoops outwards from that rock-based misty isle.

There was no puffing hole on the green shores of Devenish Island. Soon, however, the source of the ghostly whoosh came into view—two huge mute swans running along the surface of the water. Like God who was Christ, they literally walked on the soft liquid. Soon they lifted upwards and with a final effort

The author by the "reproductive cross" (see chapter 10) on Devenish Island (during World War II, 1945).

whooshed away into the mist over the bow of our dory.

For a fleeting moment, as they tucked their webbed feet against the soft down of their tail panel, I could see little drops of glistening water streaming off their webbed feet and evaporating in the morning air. There is a Japanese haiku poem that describes the power in one such drop of water.

> Let all my life of dust be cleaned by you
> On one clear evanescent drop of dew!
> —*Basho*

On that early winter morning, in the middle of World War II, I wondered why poets, such as Yeats or Wordsworth, or even the Japanese poet Basho, whom I had never heard of in 1945, could find God in little things, such as a drop of dew, while others, who seek God in economics or political power never seem to find him.

Physicists say that there is enough energy in a single drop of water to run the city of New York for a century. If it is true, I am glad we do not know how to extract such power. We have enough problems with a pound of uranium without opening the Pandora's box in a drop of water. I am older and wiser (yes, the old are wiser) and I understand such things better than I did as a twenty year old soldier on my journey to Devenish Island during World War II.

On that morning 46 years ago I began a life-long love affair with the mysterious stone structures called the round towers of Ireland.

The war had ended in victory and I was celebrating my awaited orders home by taking a break to visit the round tower at Devenish.

I do not remember when I heard about the round tower on Devenish Island. I did not at that time have a written guide to the antiquities of the Irish countryside, and I suspect that in 1945 there was not a detailed one in print. I probably heard about Devenish while sipping tea and sampling the delicious home-made Irish soda bread, beside a turf fire, in some Irish cottage along the Belleek road.

Glendalough Round Tower: The most famous round tower in all of Ireland is in the old monastic site founded by St. Kevin who died in 618. Glendalough is on a point of land between two mountain streams in the beautiful valley of Glendalough. The tower is built of mica schist, one of the most paramagnetic of all stones. Note the offset placement of the windows. Glendalough is considered by Irish scholars as the first functioning university in the west.

The abbey and round tower on Devenish Island in Lough Erne, Fermanagh, North Ireland.

During the war my knowledge of round towers of Ireland was so minimal that I actually believed the Devenish Tower was the only one in existence.

In later years as my wife and I returned again and again to Ireland we discovered many more of the elegant structures along the byways we explored. Some of the stone towers, such as the one at Drumcliff near Sligo where William Butler Yeats is buried, are only short rocky stumps; however, at least twenty-five of sixty-four remaining towers are intact. These have survived for over 1,400 years as silent guardians of Ireland's early Christian past.

Since World War II the popular literature on Irish antiquities has multiplied by leaps and bounds, as, indeed, has my own personal library on Irish subjects.

In early 1976, Eason & Sons, publishers in Dublin, printed a small pamphlet by Professor Lennox Barrow on Irish round towers. In 1979 he published his much larger study and gazetteer covering all of the known monastic towers. The gazetteer has a map of Ireland, drawn by Barrow's wife, with all the locations of towers plotted. It is a quite accurate representation and we utilized it to locate other round towers.

I spent that winter day on Devenish photographing the graceful round tower and the remains of the 12th century Augustinian priory that still survives on a low knoll behind the tower. In the graveyard is an intricately carved Gaelic cross.

I sat at the base of the tower (the high doorway was boarded up so I could not enter) and I sketched with water colors the tower, cross, and abbey. Those color sketches, plus the grainy Victorian photographs that I took with my little Foth-Derby camera, were the basis for an oil painting.

As the sun settled behind the green hills far to the west, I collected my paints and walked to the shore to meet the returning fishermen. It was around 4:00 p.m. and I had spent over two hours at the base of the tower. At times I seemed to be in a relaxed, almost pensive meditative state and yet my mind was alert to the technicalities of handling my camera and paintbrush. It was an experience similar to my climbing the

Martello tower on Ireland's Eye. It might, of course, have been the solitude of that Lough Erne Isle that induced the meditative state of my mind. As Michael Horner has quoted from Rassmussen in his book *The Way of the Shaman*, "[T]he best magic words [or thoughts—my addition] are those which come to one alone out among the mountains. These are always the most powerful in their effect. The power of solitude is great and beyond understanding."

The great biblical mystics and medieval saints always retired alone to the wilderness and solitude in order to meditate and write.

As the little dory plowed toward the distant lights of Enniskillen Quay I took a last look at the sharp pointed stone tube. It seemed to pierce the dark evening sky. Was there a special power hidden in that tower similar to what physicists say is in one drop of water? If so, can we ever understand it or even life itself?

I do not really believe that humankind will ever comprehend life, nor will the greatest scientific minds of the world be able to decipher exactly what is meant by energy—E equals mc2 notwithstanding. In the final analysis life is what each individual "feels," and what one feels involves both the intellect and the essence, or the soul, as theologians label it.

It is in the naming and classification of the ingredients of life that we somehow destroy the essence of what life really is. Science is the art of pigeonholing and fitting data. You may read these chapters of my ramblings about Ireland, and my thoughts on Ireland, all of which are my descriptions or classifications of the land and her people, and yet the Irish poet Francis Ledwidge in a few short poems gives one more insight into the mystery of life, especially in Ireland, than any book written about that country (see bibliography—*Fields of Song*).

Certainly Ledwidge understood life and its relationship to God better than scientists and theologians I have studied, including my favorite philosopher, Tielhard de Chardin who was both a scientist and a theologian. Tielhard believed that science takes one closer and closer to the omega point, that is, to God

The Ledwidge workman's cottage beside the road between Drogheda and Slane (2 miles east of Slane). It is a typical duplex workman's house at the turn of the century. It is now a museum.

and complete understanding. I agree but the problem is that we must classify and fit God to our data. Ledwidge did not, thus, he once wrote:

> I am a thought of God's now long
> Forgotten in His Mind, and desolate
> With other dreams long over, as a gate
> Singing upon the wind the anvil song,
> Song of the Spring when first He Dreamt of me
> In that old town all hills and signs that creak:
> And He remembered me as something far
> In old imaginations something weak
> With distance, like a little sparkling star
> Drowned in the lavender of evening sea.
>
> —*God's Remembrance*

Scientists are terribly afraid that life is only a cruel metaphor; that we really are not, as the catechism teaches, made in "The image and likeness of God." In fact, there may not even be a God—only a universal mechanistic clock set off by some huge explosion (of what?) called the "big bang." The result of the clock concept of science, of course, is that nature can be conquered and improved on.

Ledwidge was more humble than that for he saw himself as a tiny thought in God's mind—a unique thought left to make its own singing way, as were the town and hills of the countryside around Slane where he grew up. Is that a pessimistic view of life? Not really, for Ledwidge later adds that God *remembered* him even though he is only one tiny star among trillions. God remembers insignificant Ledwidge just as I remember the glistening beauty of water droplets that fell from a swan's toes and were lost in the "sea" of Lough Erne. The memory is distant, but *always* in my mind, stored there by the mystery of my own essence. I shall never forget those streaming drops just as God never forgot the essence of Ledwidge—his soul.

The poetry of Ledwidge tells me that Einstein missed the point with E equals mc^2. The real energy in that insignificant stream of swan water is that it was remembered by me, and perhaps, if people inherit memory as do butterflies, by my son and daughters and even their sons and daughters. It is this remembrance of things beautiful that Eugene Marias, the great South African termite naturalist, called *The Soul of the White Ant,* and I, imitating that genius, called *The Soul of the Ghost Moth.*

Some say that Ledwidge, and William Butler Yeats, were reincarnationists, but I doubt it, as they both believed in that essence, the individual soul. I would suggest that reincarnation is an egotistical concept (General Patton was certain that he had been Julius Caesar) containing as it does the seeds of false humility. The reincarnationist theory says that there is no essence of me, that I am a million other beings back through time. God allows my essence to jump from creature to creature like a pretty lass at a cocktail party.

Inherited memory, on the other hand, is a summation of every person's thoughts and is handed down by the "teacher" which lies hidden in the soul of each of our ancestors. It is the real energy of the atom and molecules of our brain, not E equals mc^2, for it passes like the words of this book from mind to mind, through the souls of our ancestors. It is mystical because it requires no printed word to reproduce and unlike books, it cannot be destroyed. The Japanese and Irish peoples seem to understand this better than most, which is why they honor (not worship) their forbears. On the other hand when professors call inherited memory sociobiology, the inheritance of cultural values, it is a vain and egotistical attempt to make of human psyche a mechanistic thing—one more attempt to circumvent God the Creator. The data search of the sociobiologists, instead of infusing a feeling of awe for creation, is one more bulwark in the materialistic concept of a clocklike mechanical universe controllable by mankind, the result of which is acid rain, stripped lands and famine in Africa, and leaky, murderous insecticide factories. Ledwidge was more truly humble and therefore wiser than we scientists, perhaps because he was born in a magic area within the shadow of the round tower of St. Cassamus the monk, which sits by the side of the road between Navan and Slane at Donaghmore.

The mysterious round tower at Donaghmore Abbey a few hundred yards from Winnie McGoona's cottage. Strangely enough this is the only round tower in Ireland without four windows around the top floor under the cap (see other tower photographs).

We once visited it. My wife and I were in the graveyard beside Donaghmore Round Tower when Mr. Dungan and his helper drove up to work in the ancient cemetery. When Norman Dungan saw us photographing the tower, he immediately postponed his work to relate some of the history of the tower. He explained that Winnie, a lady who lived in a nearby cottage, would be able to tell us far more about the tower, as she had been born 92 years ago in a nearby cottage.

As fate would have it, this Winnie had known Ledwidge when she was a little girl. Or was it fate? I am not sure I really believe in fate, perhaps God gives us a basket of fruit to choose from, called free will. At any rate, when Norman Dungan told us he would lead us to Winnie's cottage we chose the right fruit, which is never to turn down the generous assistance of the Irish.

The way to Winnie's place was a tree lined lane which opened into a clearing in which stood a two-room cottage and byre. Except that it had a slate roof, the cottage might have been a duplicate of that of Biddy Early in County Clare.

We knocked at the open door and were met by a lady with an only slightly wrinkled face, which belied her 92 years. Her name was Winifred McGoona, a name that meant nothing to us until she explained that she was the younger sister of Matty McGoona, Ledwidge's closest boyhood friend. I was unaware of the close friendship. Winnie told us that her now dead brother was a self-taught naturalist. I had already deduced that from the back wall of the cottage which was covered floor to ceiling with bookcases containing volumes of all branches of natural history, science, and philosophy. Matty was an excellent example of that strange phenomenon that extends back to the "hedge school"—a cottage philosopher. I have met such Irish in every county in Ireland and they are far wiser and better educated than many people with university degrees. Most nations can claim only a few such poets, writers, and philosophers but every other cottage in old Ireland seemed to produce at least one, including John Tyndall the scientist, and Ledwidge and Patrick Kavanagh, the poets. Both the latter, the

Winnie Callahan (left) and Winnie McGoona (right) sister of Matty who was Ledwidge's best friend, in front of the McGoona cottage.

nature poet and the agricultural poet, are modern examples of the cottage philosopher phenomenon, and because of that they are often lumped together despite the fact that Ledwidge's poetry is Victorian and in the romantic lyrical tradition, even sentimental, and Kavanagh rugged and earthy.

Winnie took a chair under the great flue of the fireplace and offered my Winnie the opposite niche. But as it was somewhat sooty from the smoking turf fire I took that spot and my wife sat in a chair in front of the huge hearth.

The soft morning sunlight streaming over the Dutch half door, the wisp of smoke from the half dead hearth fire, and the resonance of Winnie's County Meath accent set the stage for a remarkable philosophers' hour around the turf fire. We might well have been back in the 18th century.

As Winnie poured us a glass of homemade damson wine she related how Ledwidge spent his off hours at the McGoona house. Matty McGoona as a young man was a printer, and Ledwidge was first a laborer, then the foreman of a road repair gang. As a teenager Matty studied and wrote about the life history of the Giant Nebula water spider and learned on his own about the flora and fauna of the Navan-Slane countryside. How much Ledwidge learned from his friend is not known but the descriptions of nature in his poems accurately reflect, more so than the work of most poets, an accurate natural history of the Irish countryside.

Matty never married and lived to a ripe old age in the cottage with Winnie. Ledwidge's life was far more tragic. He lived to see only one of his volumes of poetry, *Songs of the Fields*, published. He was killed in action on July 31, 1917, by an artillery shell near the Franco-Belgian border in what is known as the Ypres salient. He was 30 years old.

Why Ledwidge enlisted in the British Army is difficult to determine, as he was not only an ardent nationalist, but even a leader in the revolutionary Irish "Volunteer" group. Some believe it was out of loyalty to Lord Dunsany, the gentle patron of Ledwidge, and a well-loved landlord in Meath. It was Dunsany's Royal Inniskillen Fusiliers that Ledwidge joined.

There was certainly no need for Francis Ledwidge to join up. The British never drafted the Irish, though Northern Ireland is part of Great Britain. Except that troops were stationed in Northern Ireland, it may be considered quite as neutral as Southern Ireland. All of the Irish, north and south, in both wars were strictly volunteers, of which there were many. A goodly number lost their lives at Ypres where over 136,000 German and British perished.

Lord Dunsany was transferred to a training battalion and survived the war so it is from his letters that we know so much about the life and times of Francis Ledwidge.

Two weeks before he was killed, Ledwidge, reminiscing in the trenches, wrote a nostalgic poem about Matty's fiddle, played during the evenings in the very cottage where my wife Winnie and I were sipping damson wine with Matty's sister. ["To Matty McGoona—who came one day when we were all gloomy and cheered us with sad music."] My thoughts flowed back to picture Francis Ledwidge leaning his bike against the byre wall and calling in to Matty to join him for a day along the banks of the great loop made by the Boyne River. The river is barely a mile from Ledwidge's stone and slate-roofed cottage, which we had passed on the road from Slane to Drogheda to the east. It was locked up tight but we read the beautiful plaque quoting one of Ledwidge's poems:

> He shall not hear the bittern cry
> In the wild sky, where he is lain.
> Nor voices of the sweeter birds
> Above the wailing of the rain.

Often I have heard that same cry, and also the soft calls of the rails and corncrake along the shores of the River Erne on a misty rainy morning, what the Irish call a "soft day."

The curve of the Boyne near where Ledwidge lived must have been an even more magic spot than the banks of the Erne, for in the great loop are standing stones as well as three of the largest ancient megalithic ruins in the world—Newgrange (now restored), Knowth, and Dowth. The spirits of the megalithic

Ballyshannon four miles west of Belleek lies across the north border in County Donegal and is the birthplace of the Irish poet William Allingham who wrote the most famous poem about the Erne valley "The Winding Banks of Erne." He was a close friend of Lord Alfred Tennyson.

The Rattoo tower door is about two feet above the raised arm of the author who is 6 ft. 3 inches in height. It is easy to leap up and gain entrance. The Vikings were no doubt as agile.

peoples, the first of the agriculturalists, haunt the banks of the Boyne.

Irishmen such as Matty and Francis Ledwidge recognized the magic and beauty of their countryside, as did the cottage poet-philosopher of Ballyshannon, William Allingham, who wrote a beautiful description of his river entitled *The Winding Banks of Erne.*

There are two kinds of Irishmen, those who are tied to that magic island by the beauty and essence of its countryside, and those who can hardly wait to leave it, a more materialistic breed, who are nonetheless also trapped by Ireland's essence into a constant yearning for its green pastures and rocky hills. The latter frequently return to demonstrate how smart they were to leave "a shur there's nothing for me here." The Mattys and the Ledwidges are the gentle hearts who understand the magic of life and are content to stay where the magic is strongest.

The conversation at Winnie's cottage soon turned to the nearby round tower at Donaghmore. It is one of the strangest of all the twenty-five round towers that still stands for its cap is sliced off at the top, and it has no windows in the upper floor. At the very top of almost all round towers are four windows, facing north, east, south, and west. It is for this reason, as well as the high door, that they were assumed to be bell towers, and also a refuge from Viking attacks.

The rather silly notion that, since the doorways are from fifteen to nineteen feet above the ground, the monks who built them climbed up and pulled their ladders after them, implied that the Vikings, who were sailors, could not climb or did not know how to smoke bees out of a hive. Anyone fleeing to a round tower would soon meet certain death, as indeed happened when Irish soldiers escaped the main British forces and took refuge in the elegant round tower at Ardmore on the southern coast of Ireland. The British commander could afford to wait them out, as little food or water could be stored in a round tower. When the Irish soldiers finally surrendered, they were all hanged. So much for round towers as places of refuge

The beautiful round tower on forsaken Scattery Island. The Abbey was founded by St. Senan in the 6th century and became a famed center of devotion. Earth from inside St. Senan's Bed (a stone enclosure) was said to have insecticide properties.

from Vikings.

The doorway of the round tower on Scattery Island at the mouth of the Shannon is at ground level and the monastery at Scattery is known to have been attacked by Vikings at least twice in its long history. The tower is still in perfect condition. Margaret Scanlan of Ennis, County Clare, told us that the village on Scattery Island was deserted in 1960 when the number of pupils in the school decreased to seven. As a girl, she and her family had to row a boat to mass in Cappa on the mainland every Sunday.

Not long into our morning with Winnie I began to understand where the idea that round towers were places of refuge from the Vikings originated. Winnie told us that years ago Matty had asked a dowser to dowse around the base of the tower to see if there were any secret underground passages from the base of the tower to the riverbank a few hundred yards away. [For those who do not believe in dowsing, which I have often seen work in Ireland, I recommend *The Divining Hand*, by Christopher Bird.] Sure enough, the dowser located not one but three secret passageways leading to the river. This explains how monks could climb into the tower, pull up a ladder and have enough time to disappear into the thick woods, taking all the treasures that were stored in the tower. At that time Ireland was thickly covered with hardwood and the monks who lived in the woods knew them far better than the Vikings, or even roving bands of Irish marauders.

Matty could never get help enough to dig out the secret passages as another peculiarity of Irish round towers is that the inner space below the doorway is usually filled part way up with dirt. The amount of dirt filling in the lower portions of the towers below the door's threshold is as variable as the door itself. In some the earth fills the hollow to the base of the door and in some there is no dirt at all. In Donaghmore the base interior is filled with debris to 1.60 meters below the sill of the doorway—the accumulation of ages, as the archaeologists would say. Mostly the dirt and debris are nothing but the sticks of jackdaw nests or the birds' skeletons. In some places, human bones have been found, leading to the belief that the base may have served as a burial place. More than likely they are the bones of those who foolishly did take refuge in such a confined trap and were slaughtered and buried where they fell by their attackers or companions.

There is considerable controversy as to when the towers were constructed. Many archaeologists believe that they were built in response to the Viking attacks which began around 900 A.D. Professor Lennox Barrow makes a better case for their being built between the fifth and seventh centuries, indeed, Donaghmore Abby is thought to have been founded by St. Patrick in the fifth century and turned over to his disciple St. Cassnus, the first Abbot. There is no doubt at all that the towers were bell towers for small hand bells, which were well known in ancient Ireland. The towers were ideal for calling the Abby workers and faithful from the countryside. They were also excellent lookout towers, as they extend twenty to thirty feet above the highest trees of an oak and alder hardwood forest.

Vikings approached by water and could be seen along the river hours before they attacked. Even if they attacked at night the high door allowed the monks time to collect their stored wealth and escape through a tunnel. In medieval days tunnels were known as exits of escape from entrapment. They were usually so well designed that it is almost impossible to find them. Even Magheramenagh, where our radio station was located, had an escape tunnel. Ostensibly it was a wine cellar, but it goes out for about a hundred yards and emerges into a woods. I used to watch the fox cubs at its exit where a large red fox reared her young each summer.

If round towers were watch towers, bell towers for summoning, or warning, the faithful, and also short-term holding defenses for escape through concealed tunnels, what else were they? Why are the windows so peculiarly placed, one here and one there, and hardly ever more than one at each of the four or five floors on the inside?

Each floor was reached by a wooden ladder from the floor

below. In most towers the windows are tiny—more like sighting holes for stars than windows. I do not know of a single tower in which a window lies directly above the doorway, which would be required in a defense structure.

Why are the towers tapered? Some archaeologists maintain that missiles dropped from the windows will fall down the wall and glance off at the base bowling over attackers. The taper of most towers is two to three degrees top to base. This is an extremely difficult taper to build into a tower.

I dropped a few large rocks from the top of the tower at Kildare and all fell harmlessly to the grass below with a thud. So much for the above theory. In most of the towers the walls are so thick and the windows so small that one cannot even lean out to drop anything. The only reason I succeeded at Kildare is that the cap is missing and the rim has been modified as battlements.

Another ridiculous theory concerns the dirt filling the space between the doorway and inside base of the tower. It was, according to one book, shoveled in to strengthen the base. The walls of the tower average from a meter to a meter and a half thick at the base, but even if they did need strengthening filling the inner space with dirt would just exert extreme outward pressure against the outer walls in direct contradiction to any physical principle of strengthening the walls.

At noon we left Donaghmore, first photographing Winnie McGoona with my Winnie in front of the McGoona cottage. We drove back by Francis Ledwidge's cottage but it was still locked. When we returned a few weeks later, neighbors told us we could get the key from Mr. and Mrs. Baxter, Peter and Pearl, of the local Historical Society, who lived at the top of Slane hill at the gate of ancient Slane Abby.

When we knocked at the door, the Baxters were having lunch but with gracious Irish hospitality they gave us tea to drink while they ate. Afterward, they insisted on personally giving us a tour of the Ledwidge cottage. Ledwidge's poems had been collected into *Complete Poems,* edited by Alice Curtayne. I bought a volume, and after leaving the sad little cottage I

The turf fire and fireplace in the kitchen of the Ledwidge cottage.

browsed through the verses and found these words in a poem entitled *The Rushes:*

> If they would tell me their secrets
> I could go by a hidden way
> To the rath when the moon retiring
> Dips dim horns into the grey.

At first reading, one might believe that Ledwidge was speaking of a secret tunnel from rath to the river, with the exit hidden among the rushes, as in the case of the round towers. How did Ledwidge know? He was no historian. I believe he had a second, more subtle meaning. A rath is a stone and earthen circle that rises doughnut-shaped from a stretching plain. They were the fortress sites of ancient chiefs. Today their grassy slopes are usually covered with whitethorn or huge beech and alder trees. Sometimes they are bare, but bare or tree grown, the locals call them fairy rings, for that is where fairies appear to qualified Irishmen, that is, those, like Ledwidge, with poetic souls.

Ledwidge seemed to know instinctively that such places contain a special secret energy, which is especially strong when the moon is waning ("when the moon retiring") and horn-shaped, dips down in the early morning sky.

Three a.m. to the dim light of false dawn was called by medievals the "hour of the wolf" for it was the time when strange things happened and werewolves were upon the land. It is also the time when during the waning moon the heated sunlit side of the moon is on the other side from the earth. Only the far infrared emits from the cooler dark surface earth side, however, the early morning is also lit up with a large amount of near Ultraviolet Light from airglow. Near UV is called blacklight because, when used in a discotheque to make magic for the sexes, it causes strange black flashes in our eyes as one dances about—it is a visual "drug" as potent as any weed. Astronomers hate the hour of the wolf because blacklight fogs their film through telescopes.

If the rushes, those "round-tower like" plants would tell me their secret, wrote Ledwidge, I could go to that mystic fairy place at the "hour of the wolf" and my soul would find peace.

I smiled at Winnie my lover as we drove over the hill and away from Slane. Perhaps we were as close as two souls could ever get on earth, but what of mystic Ireland? It has, after all, in song, been called "a little bit of heaven."

10

SILENT MUSIC

EVERYTHING HAS ITS SPIRITUAL MEANING, which to the literal meaning is what the soul is to the body.
—Nathaniel Hawthorn

Silent Music was what William Johnston, the Belfast mystic priest, called his book on the science of meditation. He wrote:

"When two people begin to love, they may chatter and use words a great deal; but as their intimacy grows, words become less necessary and are even superfluous. The two maintain a total and unitive silence, enjoying one another's presence in love. And the same holds true when the beloved is the mountains, the solitary wooded valleys, strange islands—silent music."

Of course *Silent Music* had not been written in 1948, but later when I read it I knew exactly what the Belfast priest meant. I had spent over three years (1946-48) mostly by myself, trotting all over Asia, and whenever I was really down all I had to do to revive was to find a close friend, or most often, a solitary wooded valley, rocky hill, deserted temple, or even more strangely a special stone to sit against—like the one on Breesy Hill in the Pullans.

It was not always, however, the stones of wilderness desert, or wild windswept moorlands, that played their silent music to my soul, it was as often a stained glass cathedral, ancient stone ring, Irish cottage, or even more strangely some graceful round tower. In fact, I believe to this day that where stone is concerned, it is the Irish Round Tower that sings loudest to my being.

The Irish round towers are unique among religious structures of the world. The fact that they are a life (phallic) symbol can not be doubted in view of the beautiful *reproductive* Celtic Cross on Devenish Island (see page 152).

I seemed to have a hierarchy of silent music, with family and good friends at the top, then birds, then insects and, finally rocks and Irish towers.[1]

Among friends are those from whom we are never really separated no matter how far away we travel. I have a few who go back to my boyhood days and two that I picked up on my round the world excursion. Then there were all my dogs, though I include them with humans. Dogs, I know for certain, have souls, for they understand silent music even better than mystic people.

I have had birds, falcons and parrots, that would rather sit on my fist or shoulder than on their favorite perch—we were friends. My African Grey parrot would open his own cage door and fly down to my writing desk and climb on to my shoulder. Birds know by instinct where the magic spots are.

I have read whole volumes about the magic spots called chakras on the human body and there is never so much as a men-

1 Since my pack trip with grandchildren and daughter across the high Sierras, plus "Blackjack," a mule, I could add mules to that list. For a real "high" hug an old mule around the neck.

The ancient Irish monks did not vulgarize sex and reproduction, but symbolically worked their love for male and female into the Christian religion. Left—when the sun is right, shadows create an erect pallus. The three braids that connect to the man's sex organs and head probably represent the strength (braided cords) of the Trinity. Right—the side of the cross is a female reproductive figure with head emerging from the vagina and a braided umbilical cord connected to what is obviously the uterus and two ovaries. From the side, the stem of the cross takes the form (hips) of a female figure. On the back can be seen the worn away figure of a crucifix. In later centuries, Puritanism reared its ugly head and made sex an evil process leading to vulgarization of the mystery by Protestants and Catholics alike. Nowhere else have I seen a cross so obviously sexual as this high cross on Devenish. The round towers, besides being watch towers and energy antennas are also, without doubt, phallic symbols.

ras on the human body and there is never so much as a mention of the space between the head and the arm called the shoulder. Every mother knows, that it is not the chakras but the left shoulder where the silent music resides. If mothers, babies, and parrots know where to perch or lay their heads, I wonder why books on chakras do not agree?

I can understand magic places around living bodies. Most scientists will admit that all living things have an aura, even if we do not understand how it works or what it is—but rocks? There are almost as many theories and books about the human aura as there are kind hearted mystics and experts to write them, but this may well be the first book on the *aura of rocks*.

The ancient Egyptians knew about both the aura of the human body and the silent force in rocks. They had two separate hieroglyphics for stone, one was an oblong brick and stood for limestone types of rock that are diamagnetic, or very weakly paramagnetic, and a second hieroglyphic with lines through it that stood for granite, porphyry, basalt and other highly paramagnetic stone.

What are these para and diamagnetic forces that I mentioned in earlier chapters? They are two very important weak physical forces that are listed for most organic and inorganic substances in the physical handbook. In general, living systems animals and plants, except for the oxygen breathed, are diamagnetic, while stones and minerals can be either diamagnetic like most limestones or paramagnetic like most stone of volcanic origin.

Where did the Irish learn of these forces? Perhaps from the Druid priests, perhaps from the Egyptians, for there is considerable evidence that the Egyptians reached Ireland. Egyptian jewelry has been found in Ireland and Ireland was once called Scotus (Scotland), the name of a known Egyptian princess (see *The Sphinx and the Megaliths* by John Ivimy, bibliography).

Early in the nineteenth century, natural philosophers, foremost among them the two good friends Michael Faraday, the Englishman, and John Tyndall, the Irishman, discovered the yin and yang of the male and female of nature—opposite weak forces: diamagnetism and paramagnetism. John Tyndall found that if he took little pieces of clean wood from over thirty species of trees and suspended them from a thread, the strips of wood would move away from a strong magnet (negative reaction). Faraday called this weak repelling response diamagnetism. Natural philosophers, mainly in Germany, also found that certain minerals were highly attracted to a strong magnet. They called this magnetic susceptibility paramagnetism (positive reaction).

I experimented with soil types in glass tubes and found that good fertile soil is highly paramagnetic. Viable soil always comes from eroded volcanic rock. Volcanic rock is far more likely to be strongly paramagnetic than sedimentary rock such as limestone.

In the physics dictionary *paramagnetism* is technically defined as; "an assembly of magnetic dipoles that have random orientation. A dipole is a substance that has a + at one end and a - at the other. If a strong magnetic force (2000 gauss magnet) is brought near a paramagnetic substance the magnet aligns the little magnet dipoles in the atoms and the newly aligned dipole field reacts by attraction to the strong magnet.

Diamagnetism is defined as "a negative susceptibility by a substance to a strong magnet." If a strong magnetic field is applied to such a substance its electrons change their orbits and velocities so as to oppose (move away) from the applied magnetic field. Most organic substances are diamagnetic. Diamagnetism is a very weak force, so weak that even though it may be in molecular mixture it is often masked by the stronger paramagnetism in the mixture. Solids, liquids and gasses can each be either paramagnetic (male) or diamagnetic (female). Oxygen is one of the strongest paramagnetic gasses that exists in nature so that blood cells are highly paramagnetic, therefore the human body, as long as it breathes, is slightly paramagnetic as it takes in a highly paramagnetic substance oxygen. If it dies, however, and no longer takes in oxygen, it becomes totally diamagnetic like Tyndall's wooden strips.

All plant life, since it gives off oxygen instead of taking oxygen in, is diamagnetic. Now dear readers, you see why I call

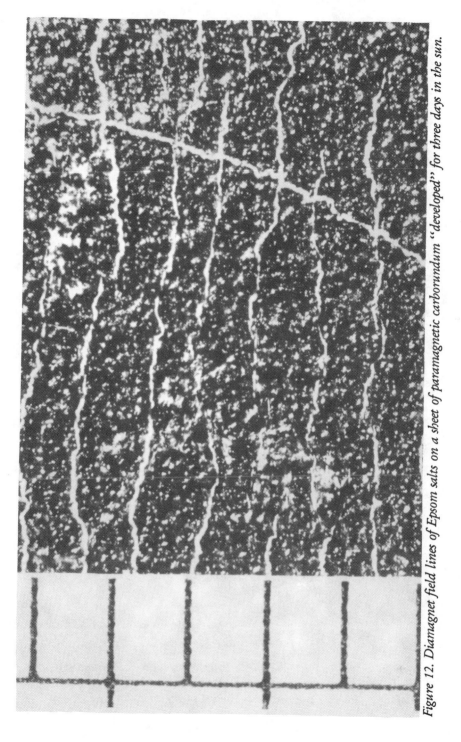

Figure 12. Diamagnet field lines of Epsom salts on a sheet of paramagnetic carborundum "developed" for three days in the sun.

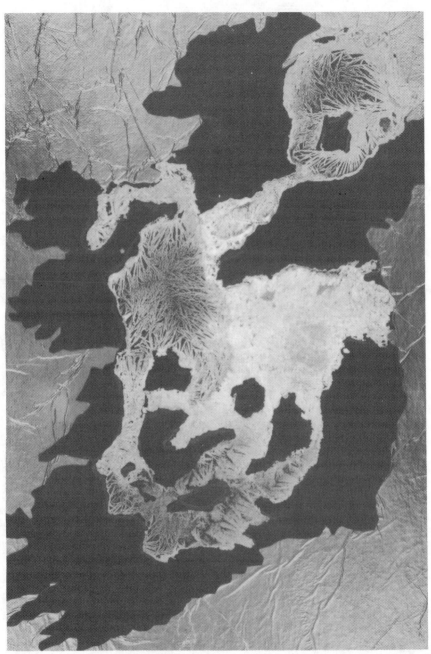

Figure 13. A diamagnetic (Epsom salts) paramagnetic (carborundum) map of Ireland. The greatest concentration of salts is between the contours that mark the rim of the mountains and covers the central plains.

paramagnetism and diamagnetism the yin and yang of life. The physics handbook has two large tables of measurements of these opposite forces. One table lists organic molecules (mostly diamagnetic) and the other inorganic atoms (mostly paramagnetic). I show a few of my own earliest (1975) measurements in table III, page 187.

There is now a more exact instrument used for computing these measurements. The MS2 Magnetic Susceptibility meter (Barington Instititute Ltd. Charbury, Oxford, 0X73PQ, England) which reads out on a digital meter in c.g.s. units. c.g.s. stands for *centimeters, grams,* per *second,* and measures the weight of the sample in grams, that moves one centimeter distance in one second. It is an excellent instrument and has been on the market for several years. Measurements made with this instrument correlate very closely with my own measurements.

My early measurements were made by hanging the rock from a thread and measuring how far it could be attracted to a 2000 gauss magnet. My readers will note that this crude early method gave excellent ratios of the strength of one rock against another (+ or -). These early comparisons demonstrated that all round towers were highly paramagnetic and thus are magnetic antennas.

Strangely enough, although physicists have spent years measuring these forces, and utilizing them to explain theoretical atomic forces, nowhere in the scientific literature has anyone, chemist, physicist, or biologist, asked what do these two opposite forces mean to life? In other words as Mark Twain once said "everybody talks about the weather but nobody does anything about it."

Now one would think that since John Tyndall proved plant life to be diamagnetic and since *all* volcanic rock is highly paramagnetic, and soil is eroded rocks, that soil scientists would be experimenting with these two forces, yet I have been unable to find the words even mentioned in over fifty soil books, *Soil Physics* (fourth edition, 1972) John Wiley & Sons, Publisher (see bibliography) does not even mention the two forces one single time. The same with the book *Soils in Construction* (second edi-

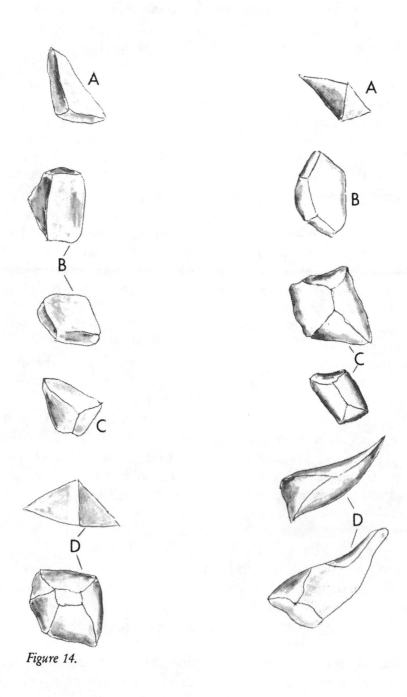

Figure 14.

tion, 1980) John Wiley & Sons, Publisher (see bibliography). One would also think since concrete (limestone) and clay (adobe) are *highly* paramagnetic and water (H_2O) is, like plant life, diamagnetic that the construction engineers would also be interested in these two opposite forces. After all, nothing happens in the growing of crops or the hardening of concrete until these opposite yin and yang forces join together.

The yin and yang of rocks and plant life is easily demonstrated. If one takes a sheet of plastic coated carborundum (the kind made for polishing metals) and lays it flat in a shallow pan of water with six to twelve tablespoons of epsom salts (magnesium sulphate) dissolved in the water, after a few days in the sun the water will evaporate. On examination the sheet of carborundum will show fine lines of crystallized epsom salt that have appeared as if by magic on the surface of the carborundum (figure 12, page 181). The lines will always run from side to side, about one mm apart, across the width of the paper. Epsom salt is diamagnetic which is the opposite force from the paramagnetism of the carborundum. Incidentally the first radio receivers called a crystal set utilized a crystal carborundum for the detector.

Another experiment is to obtain an accurate geological survey map of a favorite area and with a hard steel stylus (or point of nail) trace and cut the contours into a sheet of carborundum sandpaper. If the sheet is then immersed in a shallow pan of epsom salts, and the solution is allowed to evaporate, in a few days an energy map of growing force lines appears (figure 13, page 182). The lines of force will concentrate in the most energetic spots on the carborundum map. Even more strangely if the map is put aside and observed for six months or a year, the crystals of diamagnetic epsom salts will begin to grow and produce little hills and mountains (figure 13, page 182). The contour forms will grow like magic forming an accurate template of the real landform. Since the crystals will reproduce and grow, who is to say that the forces of rock and soil are not living! We may understand then that by very simple fifty cent experiments we can demonstrate for ourselves the power of these weak forces to grow and accumulate energy.

If one is willing to invest a few dollars in a very small but powerful (2000 gauss) magnet then it is easily demonstrated that these weak magnetic antenna forces are different from strong magnetism and that the paramagnetic force depends on the shape of the rock. In other words you can shape a rock to give off energy just like you can shape a radio transmitter or radar antenna to amplify and throw off radio energy more efficiently. The contour lines cut into carborundum will concentrate, or make coherent, the paramagnetic force field in the same manner as cutting rocks to specify shapes amplifies the paramagnetic energy.

Paramagnetism, then, is not magnetism, but the ability of a substance, property if you will, to collect, tune into, resonate to, the magnetism of the earth and sun (sun spots). Earth and sun magnetism (cosmic) is an alternating (changing) force like AC current, not steady and fixed like a magnet (DC).

If one piles up various forms of magnetic substances, like screws, paper clips, pins etc., their shape will make no difference, for the magnet will attract all of them. If, however, one takes a hammer and crushes a red clay flower pot, pieces of granite, basalt, etc. and touches the magnet to the pile of grains, two piles will emerge, the one that clings to the magnet and a second that is left behind (see Table III, page 187). If the rock is all diamagnetic you can not pick up any of it. If it is weakly paramagnetic only a few particles of the crushed rock will stick to the magnet. If it is highly paramagnetic, like pink granite or basalt, most of it will stick to the magnet. If one then takes the pile of ground up rock, or clay flower pot, which did not stick to the magnet and smashes it again two piles will separate out once more—one that sticks and one that does not! This goes on and on *ad infinitum* as the crushed shapes change.

Look at the two piles under a good magnifying glass. There will be considerable variation in shape but in general the ones that cling *will be* conical or flat conical (figure 14, page 184, left A), or flat stonehenge—gravestone shape (B), irregular pyramidal (C), or regular pyramidal (D). Pieces that will not be

186 *Nature's Silent Music*

TABLE III
IRISH ROCK MAGNETIC FORCES*

Type	Magnetism	Force (1 to 5)
1 Basalt	Para	+5
2 Green Slate	Para	+4
3 Old Red Sandstone	Para	+/−3
4 Chalcopyrite (Copper Ore)	Para	+3
5 Connemara Marble	Para	+/−3
6 White Granite	Para	+2
7 Carboniferous Limestone	Para or Dia	+2 or -1
8 Dolomite (Limestone)	Para or Dia	+1 or -1
9 Quartz	Para or Dia	+1 or -1
10 Sphaherite (Zinc Sulfide)	Para or Dia	+1 or -1
11 Barite (Barium Sulfate)	Para or Dia	+1 or -1
12 Galena (Lead Sulfide)	Dia	-1

* The force is my own measurement in millimeters (mm) distance that 1 gram of paramagnetic rock will move to a 2000 gauss magnet. Certain rock types may by para- or diamagnetic depending on where they originate. Burran limestone is mainly diamagnetic. Diamagnetism is an extremely weak force and never moves away from a 2000 gauss magnet more than one mm (-1). We see from this table that the green slate of Yeat's Ballylee Castle (page 33) is extremely paramagnetic (energetic).

attracted are (right) elongated triangular (A), flat odd shaped (B), Irish roof cottage shaped (C), curved or odd shaped (D). We may understand then from these two experiments that, although the stone or clay is still stone or clay, as far as paramagnetic material is concerned, some forms transmitted the field of energy outwards, as in the attractant shapes (left) or inwards (does not transmit) as in the forms that can not be picked up (right).

A stone that is Irish cottage shaped is still paramagnetic, but is not usually attracted to a magnet, therefore the force is more than likely directed inwards and not outwards as in the stonehenge and pyramid shapes. We see then that such a house must focus a lot of energetic paramagnetic forces into the bodies of those who dwell therein.

But what about the diamagnetic straw roof? In fact, the roof is not completely straw, the base of all thatched roofs is heavy grass sod (paramagnetic soil) rolled over the wooden beams that support the thatch, allowing thatchers to use wooden staples to hold the thatch in place. A thatched roofed Irish cottage, then, is built of paramagnetic stone walls and a roof of paramagnetic sod and soil with a thick layer of diamagnetic straw insulating it from any magnetic force from above, e.g. moon, sun, etc. All the life energizing paramagnetic force is directed inward.

The Irish rural folk that during World War II lived along the Erne River were hard working and energetic, yet despite their labors they did not wear themselves out, or get sick very often.

Their excellent health was noted by the writer Isaac Bulter (1749) in his *A Journey to Lough Derg*. Writing about the Erne valley he said: "This tract of country is healthier in an uncommon degree, so that physician or even apothecary would find it impossible to support himself by his practice."

What about Martello towers and round towers? Model towers of carborundum, scaled to the same dimensions as Irish round towers and dipped in epsom salts, demonstrate the same energy rings, after drying, as a flat sheet of paper. They will circle the tower about one mm apart from top to bottom. A person clinging to a large rock paramagnetic surface would absorb the paramagnetic force lines into his body.

Paramagnetism in round towers is evident from the plant growth on many of them. The mortar must be particularly paramagnetic since it was, according to tradition mixed in oxen blood (iron doped).

A small model tower placed in the center of a flower pot with seedling radishes planted around the base of the tower will stimulate the seedlings to grow many more fine rootlets than a pot with the same soil but no central round tower. The ancient Irish monks were known to grow the best gardens in the country. Their gardens were planted inside the walled compound as where the monastery round tower stood.

The morphology of a cross section of the Irish hedgerow is an excellent example of the intersection of the paramagnetic force of stone with the diamagnetic force of the water and the organic matter of plants—the roots always grow *sturdier* and more numerous among the stone.

The Burran I have spoken of in County Clare is barren limestone rock, cracked and eroded by time. Very little grows there. Burran stone is one of the few stones I have tested that is completely diamagnetic (Table III, page 187) and diamagnetic plant life does not grow from diamagnetic stone. All over Ireland the stone ditches (walls) are covered with hedgerow plants. Most stone walls, even where there are no bushy shrubs, are at least covered with ivy—not so the stone fences of Galway, the Aran Islands, and other parts of Connemara. The rock walls of Connemara are mostly bare because they are built of diamagnetic limestone.

The subsurface limestone of central Ireland is more than likely of a diamagnetic form. Over the centuries it has been completely covered with a thick layer of eroded stone—paramagnetic soil. The Burran is a special coastal region where the subsurface diamagnetic limestone thrusts up above the covering mantle of earth. The eroded rock soil was channeled into the central basin of Ireland by the hard red sandstone ridges (see Chapter 3).

Ireland takes the shape of a highly paramagnetic ring of volcanic rock surrounding a base of diamagnetic limestone covered with a deep layer of eroded volcanic paramagnetic soil. Of course, good soil is not just paramagnetic (yin); it must also have diamagnetic force (yang), which includes both water and decaying organic matter. Consider, however, that with diamagnetic limestone dissolving from below and mixing with the water and organic remainder of crops or forest plants, in Ireland nature produced both a strong yin and an equally strong diamagnetic mixture of yang. Strong, but slow and steady male and female forces join together in Ireland to produce healthy offspring of forests and crops, and also, over most of Ireland, the very best grass in the world for healthy livestock.

What about the effect of these opposite forces on populations? Most of Northern Ireland sits directly on a subsurface of highly volcanic basalt. I am not the first observer to point out that the people of Belfast are more energetic than those of Dublin, which has a limestone subsurface, or that the people of granite-based New York are more energetic than those of limestone subsurfaced Florida where I live. I am not saying that Dubliners, or my fellow Floridians, are lazy, only that although they work as hard as anybody, there is a certain subtle laid back aura evident in limestone lowland people, and that the exact opposite energetic aura is evident in such places as Belfast and New York City. If, in any highly paramagnetic region of the world—Belfast, New York, Lebanon, certain mountain regions such as North Vietnam (not south, which is subsurface limestone)—the people are forced into unemployment, they are very likely to start beating on each other.

If sociologists were more aware of the effect of such natural low energies on a people, they would include what I call sociophysics in their studies. The biggest mistake that modern science makes is that it breaks things apart for study but seldom puts them back together to learn how the *system* works. We hear a lot about ecological systems these days but there are very few ecologically oriented physicists, chemists, sociologists, or even agriculturists. In fact, it is the agriculturists who preach high energy, destructive, quick-profit agriculture at the expense of the small family farmer (a truly first-class system ecologist), who are destroying the environment.

The land on the Burran and the Aran Isles was efficiently farmed by Celtic farm families who learned how to fill the little valleys, cracks and crevices, and higher flatlands between the diamagnetic rocks with a mixture of soil made from organic seaweed and eroded volcanic dirt. Of course, most of the dirt was already there, the result of centuries of windblown erosion from surrounding mountains, carried down and trapped in the cracks and crevices of the Burran stone. This occurred over the eons in much the same manner that the tributaries and streams, and the glaciers of the Pullan, carried eroded quartz, granite, schist, and blue mountain limestone, and the organic remains of Lough Erne, into the fertile valley of the Erne, that magic energy spot where I lived from 1944 to 1946, and where first I stumbled on the upland huts of the booley girls.

There is no question that Biddy Early understood the power in stone, and that she was a booley girl in her teens.

Meda Ryan in her book, *Biddy Early—the Wise Woman of Clare*, states that Biddy had a favorite stone in the corner of her parents' little farm where she sat and communed with the fairies. The Early farm would have been too small to save enough hay for the winter to feed the two cows they owned. To the north of the family farm lie high bogland hills where she would summer the cows. Today there is a modern high energy emitter spoiling the view of those purple-green mountains. It is called a TV antenna.

Biddy and the other girls of the Feakle area must have spent at least a part of their summers on those mountainsides, leaving behind the lowland summer grass so it could be saved to carry over for the cows during the winter months.

When Biddy's parents died she became a nomad-survivor of the open roads. Her booleying days on rocky hillsides and under blue skies must have put her in touch with the low energy "fairy forces" that kept her alive in the harshness and seclusion of her young maidenhood. Surrounded by rocks and heather with only the lonely cry of the mountain curlew to fill her days, she listened to the silent music of the hills and from that music developed her marvelous curative powers, utilizing both herbs, and perhaps also a knowledge of rock generated healing power. Healing stones and rock pillars are found all over Ireland.

If the paramagnetism of stone and oxygen are important to growth and to nature's system of life, what of the gentle breeze? We now understand that just as breath is the spirit of human life, the flow of the zephyrs of the green Erne valley and the lonely Pullan are the life spirits of the earth. It is the gentle paramagnetic (oxygen) breezes that set the dancing leaves of the hedgerows vibrating. It is also those Pullan zephyrs that flow across the stone walls and yellow thatches of the Erne valley cottages, vibrating each little straw of thatch and colliding with each little bump of the wall so that the tiny molecules from the heather, the grass, the whitethorn, and the cow byre are stimulated by the pressure waves (called phonons) to emit energies of life. The people of the Irish cottages and Erne valley absorb these life frequencies into their bodies, which is why in days gone by they did not need doctors.

What has gone wrong with the agriculture or worse what is happening to the world? Man in his technological arrogance is tearing down the Irish cottages of the world and dumping tons of poison into the yin and yang of the soil. He is ripping out the dancing hedgerows, cutting down the rowan trees, and doping the gentle breezes, "the spirit of life" with polluting chemicals that form acid rain.

I believe that John Cardinal Wright, in his introduction to *The Delicate Creation* by Christopher Derrick, outlined the modern mentality that leads to the type of corporate farming that destroys our living soil. He wrote:

> "On the one hand we can make the mistake of the materialist: We can suppose that this world, and the observable universe beyond it constitutes the whole of reality and is therefore the only thing that we need to worry about. In general, religious people at least are aware of this mistake and on their guard against it. But they may be less keenly

on their guard against the converse mistake—the idea (often cherished by deeply spiritual people) that the material universe is somehow a contemptible thing, unimportant at the best, positively evil at the worst."

It is, as Nathaniel Hawthorn indicated, a deadly error to attempt to separate in our intellect, the spiritual from the physical—body and soul are one, as is life and the paramagnetic soil.

EPILOGUE
AN AFTERGLOW

THE MOST IMPORTANT principle I wish to impart is that of treating rocks, stone and even the soil as antenna collectors of magnetic energy waves. Round towers are conical magnetic antennas, rocks are antennas according to their shape, and even the soil is a huge flat ground antenna if it is composed of a sufficient amount of the volcanic paramagnetic eroded rock. THAT IS PRECISELY WHY, IF A FARMER LETS THIS HUGE SOIL ANTENNA ERODE OR BLOW AWAY, HE LOSES ONE OF THE MOST IMPORTANT FORCES ON HIS FARM. Organic matter which is *diamagnetic* is the other force and just as necessary. Another important factor is the diamagnetic organic matter *stores* water, but it is the opposite force the volcanic paramagnetism that controls the evaporation of water.

If you ever drive from El Paso in West Texas to San Antonio, leave Interstate 10 at Kent (about 140 miles) and turn south on Texas highway 118 where it cuts through the Davis Mountains of Texas and hits Texas highway 90 after passing through Fort Davis, a famous army post during the Apache Wars. It is now a fascinating National Historic Site. The little known all black 24th and 25th Infantry, foot soldiers, defeated the Apache tribes from there.

The Davis Mountains, which are of tertiary volcanic basalt lava, are high enough to catch a little rain, not much, but they control the evaporation in their *highly* paramagnetic innards. The low range is thus a paradise of water and green in a vast upland plane of desert. That is of course why the cavalry built a fort there—*water* and *garden* for food.

The white man and the water slurping horse could not stand up in the desert, so the black 24th and 25th infantry (called Buffalo soldiers) outwalked (chased) the Apache across the Mexican border and they never came back. The history of the American black is as distorted as the information fed farmers by corporate agriculture.

My meter for measuring the paramagnetic force is called a magnetic *susceptibility* meter. An antenna is a form of a material substance, e.g. metal and radio, that is *susceptible* to a frequency, one can also say "collects a frequency," "resonates" to a frequency, "tunes to" a frequency, etc., etc. Paramagnetic "flat" soil then is "tuned to," "resonates to," is *susceptible* to magnetic frequencies from the sun and cosmos.

Soil is not *magnetic*. It does not *store* the energy like a magnet. A magnet is analogous to a battery which stores an electric field instead of a magnetic field as does a magnet.

It is true that magnetic energy will stimulate root growth, but using magnets for stimulating plants is not practical, it is analogous to bringing nuclear energy, the sun, to earth to run electric plants. Anyone who still thinks that practical probably will not like this book.

I am certain that common sense farmers will come to understand that restoring this paramagnetic energy collector to their soil with truck loads of volcanic gravel or ash plowed into their organic filled earth is a much more practical idea. For small gardeners it is extremely practical as is the building of six foot miniature round towers in a garden, see *Secrets of the Soil* (chapter 15 and 21), by Bird and Tompkins.

It is much better to do all things as God intended by capturing the magnetic frequencies from the sun with the paramagnetic soil or round tower antenna. In one sentence MAKE YOUR SOIL SUSCEPTIBLE (RESONATE) TO THE SUN, do not, with magnets, try to bring the sun to earth.

A truly believing materialist will depreciate this book. His or her arrogance will take over. It is the same godless arrogance that calls Biddy Early a superstitious witch, and scorns her low energy cures. One may have less than a 30% cure rate of cancer with modern technological medicine but a 30% cure rate with low energy homeopathic herbs makes the healer a fraud to be persecuted by the FDA. Laying on of hands, despite the fact millions worship the greatest hand layer on'er of all times, Christ, is scorned by modern man, as is the gentle donkey cure that honest people say sometimes works. The donkey healers are frauds and charlatans.

It is true that many times these country cures do not work, just as many times chemotherapy and other high energy cancer cures do not work, but that is because we do not understand enough about the low energy coherent frequencies that control life. Christ was always successful for being divine, like Mohammed and Buddah and other sons of God, He completely understood such systems. Scientists will probably never completely understand but if the great philosopher Teilhard de Chardin is correct the study of nature (science) will lead us closer and closer to God.

When Jack Perkins, the well known NBC commentator, once asked me on NBC Magazine if I believed in miracles, I replied with a question "Jack do you believe in life?" "Of course" was his reply. The program ended with my "Well!"

A great American Indian made the connection between God and our soul's desire to understand nature in two short sentences. The genius of such poets is that they can summarize an entire life and give a meaningful view of nature in a few lines.

The Indian Blackfoot Chief Crowfoot in his dying hours asked; "What is Life?" He answered the question himself;

"It is the flash of a firefly in the night. It is the breath of a buffalo in the winter time. It is the little shadow which runs across the grass and loses itself in the sunset."

The flash of a firefly is room temperature cool and light. The breath of a buffalo is paramagnetic oxygen. A little shadow that runs across the grass is caused by movement or vibrations. How did an Indian brave understand by instinct what I, or so called civilized modern, took forty years to discover? Perhaps it is because the so called savage Indian could really speak to God while moderns, such as I, speak only to our own egos. Let us hope we turn back to God before it is too late.

ANNOTATED BIBLIOGRAPHY

The books listed are those I enjoyed and that also gave me an insight into some aspect of the culture, history, wildlife and geography of Ireland. Many are written in a popular vein, some are the experience of travelers overcome with a spell by that fair land similar to my own. A few are masterpieces of scholarship and readability e.g., *Irish Heritage* by E.E. Evans, *The Irish Landscape* by Frank Mitchell and *The Flowering of Ireland* by Katherine Scherman. A few evenings browsing through these classics will hook one forever on the green hills and hospitable people of that small island.

The first few books are of specialized subjects, such as the one on climate, that are handy for the serious traveler in Ireland. The rest are listed in the same general format as the book, under *form, people,* and *magic*, with a last section for the naturalists among my readers.

Finally I do not document my own scientific work in this

book. A list of my scientific papers up to 1975 is included in *Tuning In To Nature*. More recent papers may be found listed in scientific abstract journals such as *Biological Abstract* and *Applied Optics* indexes. All were published, scholars take notice, in reputable scientific journals and number well over one hundred.

BOOKS THAT CONTAIN SOME INFORMATION ABOUT IRELAND AND ARE QUOTED IN THIS TEXT

The Round Towers of Ireland. Barrow, George Lennox. Dublin: The Academy Press, 1979. A very fine gazetteer and study of all of the round towers of Ireland.

The Divining Hand. Bird, Christopher. New York: E.P. Dutton, 1979. The best ever history and explanation of techniques of divining for water or whatever by one of the authors of the *Secret Life of Plants* and *Secrets of the Soil*.

The Stone Circles of the British Isles. Burl, Aubrey. New Haven and London: Yale University Press, 1979. A fine gazetteer of all the stone circles of England and Ireland.

Tuning In To Nature—Solar Energy, Infrared Radiation and the Insect Communication System. Callahan, P.S. Old Greenwich, Conn: The Devin-Adair Co., 1975. Covers my own work with insects and the infrared portion of the spectrum.

Ancient Mysteries, Modern Visions—The Magnetic Life of Agriculture. Callahan, P.S. Kansas City: Acres U.S.A., 1984. Contains my own studies on the round towers of Ireland and the pyramids in Egypt, and explains how they relate to the paramagnetism of soil and stone.

Irish Heritage. Evans, E. Estyn. Dundalk: West Tempest, Dundalgan Press, 1967. A classic work on rural Ireland reprinted 9 times—for an overview of all agricultural Ireland. The best book of all.

An Irish Beast Book. Fairley, J.S. Belfast: Blackstaff Press, 1975. An excellent natural history of Ireland's mammalian wildlife.

B.C. America. Fell, Berry. New York: Pocket Books, Simon and Schuster, 1976. A fascinating account of all the archaeological evidences that America has been visited many times by ancient peoples.

Where the Erne and Drowes Meet the Sea. Gallachair, P.O. Ballyshannon, Donegal: Donegal Democrat Ltd., 1961. The only small history of the Ballyshannon-Belleek area described as "fragments of a patrician parish."

The Parish of Cairn. Gallachair, P.O. (No publisher given). The only history of the Belleek area. Taken from the writings Cannon Kenna over half a century ago.

Paddy's Lament. Gallagher, Thomas. New York: Harcourt Brace Jovanovich Publ., 1982. An excellent history of the Irish famine of 1846-47.

The Irish Landscape. Mitchell, Frank. St James Place London: Collins, 1976. No Irish library would be complete without this scholarly work on the land form of Ireland.

Hedges. Pollard, E. M. D. Hooper and N.W. Moore. St. James Place London: Collins, 1977. An excellent book on the ecology of hedgerows in England.

Antiquities of the Irish Countryside. O'Riordain, Sean P. London: Methuen & Co., 1976. An excellent guide (pocket size) to the dolmans, megalithic tombs, stone circles and other ancient stone monuments of ancient Ireland.

Biddy Early, the Wise Woman of Clare. Ryan, Meda. Dublin: The Mercier Press, 1978. An excellent paperback biography of Biddy Early.

The Flowering of Ireland—Saints, Scholars and Kings. Scherman, Katharine. Boston-Toronto: Little, Brown & Co., 1981. A thorough and highly readable history of Ireland from prehistoric times through Christianity.

GENERAL BOOKS THAT MAY BE HELPFUL

A Dictionary of Irish Biography. Boylan, Henry. Dublin: Gill & Macmillan Ltd., 1979. Short biographical sketches of Ireland's greats; poets, scientists, politicians, etc.

Dublin. Clarke, Desmond. London: B.T. Batsford Ltd., 1977. A readable overview of the beautiful city of Dublin.

An Atlas of Irish History. Edwards, Ruth Dudley. London: Methuen & Co., 2nd ed., 1982. An excellent historical geography of Ireland. The maps are beautifully done.

Ireland, Blue Guide. ed. Ian Robertson. New York: Rand Mc-
Nally & Co., 1979. By far the most comprehensive pocket
guide of Ireland. Historically accurate.

Irish Agriculture in a Changing World. ed. I.F. Baillie & S.J.
Sheehy. Edinburgh: Oliver & Biyde, 1971. A good summary
about changing agriculture, for better and worse, in modern
Ireland.

Irish-English, English-Irish Dictionary. Ireland: The Talbot Press,
1976. A nice little pocket source for common Gaelic words,
useful for nameplaces.

The Surnames of Ireland. MacLysaght, Edward. Dublin: Irish
Academic Press, 1978. An excellent little volume on the
derivation and spelling of Irish surnames.

Topography of Ireland, a translation of the Latin by the Welsh
born bishop Giraldus de Barri, called *Cambrensis* (1146-
1223). O'Meara, John J. Dundalk: Dundalgan Press, 1951. A
geographic, mystic and somewhat biased view of what the
medieval Britton thought of Ireland. Strangely enough it is
organized almost exactly as this book into three sections. I.
Geography and Natural History (Form) II. Wonders and
Miracles (Magic), and III. The Inhabitants (People).

Promise and Performance—Irish Environmental Policies Analysed.
ed. J. Blackwell and F.J. Convery. Dublin: University College, 1983. An overall ecological view of modern problems.

Prospect of Fermanagh. Rogers, Mary. Enniskillen: Watergate
Press, 1982. A very fine pocket size guide to Fermanagh
County and the Erne valley where Belleek is located. Contains a beautiful section on Devenish Island. Places of interest are referred to the one Inch Ordinance Survey Maps (see
section IV).

The Climate of Ireland. Rohan, P.K. Dublin: The Stationary Office, Gov. Post Office, 1975. Check the tables to guess
about warm sunny days in Ireland.

IRELAND'S FORM

A World of Stone. ed. O'Sullivan, Paul. Dublin: O'Brien Educational Press, 1972. An excellent textbook geography of the rocky Aran Islands.

The Personality of Ireland, Habitat, Heritage and History. Evans, E. Estyn. Cambridge, England: The University Press, 1973. A classic series of lectures (small book) on the heritage and form of Ireland by the great folklorist and geographer (see also *Irish Heritage*).

Ireland. Freeman, T.W. London: Methuen & Co., 1972. The best book on the geography of Ireland.

The Corrib Country. Hayward, Richard. Dundalk: Dundalgan Press, 1968. A romantic and artistic look at the beautiful countryside around Lough Corrib in the west of Ireland. The sepia wash drawings by J.H. Craig are little gems.

Celtic Monasticism. Huges, Kathleen & Ann Hamlin. New York: The Seabury Press. A nice book on the structure, location and organization of early Irish Monasteries.

J.M. Synge's guide to Aran Islands. ed. Ruth W. Shaw. Old Greenwich, Conn: The Devin-Adair Co., 1975. A beautiful guide to the rocky Aran Islands based on the writing of the great Irish playwrite J.M. Synge.

Island Outpost of Europe. Lavelle, Des. Skellig. Dublin: The O'-Brien Press, 1976. A beautiful book on the form, geography and natural history of Skellig Island, one of the most fascinating volcanic rocks on the coast of Ireland.

Irish Churches & Monastic Buildings. Leask, Harold G. Dundalk: Dundalgan Press, 1966. Excellent book on the architecture of monastic buildings in Ireland.

Irish Castles and Castellated Houses. Leask, Harold G. Dundalk: Dundalgan Press, 1977. The best ever small book on the physical structure and history of Irish castles and tower-houses. A nice cover map of where each is located.

The Monasteries of Ireland. Pochin Mould, Daphne D.C. London: B.T. Batsford Ltd., 1976. An excellent scholarly history of Irish monasteries.

The Mountains of Ireland. Pochin, Mould, Daphne D.C. Dublin: Gill & Macmillan, second ed., 1976. A very readable geography of the 1% of Ireland that makes up the mountainous, rocky rim of pastoral Ireland. A must for mountaineering.

The Development of the Irish Town. ed. R.A. Butlin. London: Croom Helm, 1977. Traces the history of urban settlement from Pre-Norman Ireland to the nineteenth century.

Early Ireland—A Field Guide. Weir, Anthony. Belfast: Blackstaff Press, 1980. A must for anyone interested in the ancient stone structures of Ireland. Structures are listed by county alphabetically and located on one inch Ordinance Survey Map.

IRELAND'S PEOPLE

The Irish Countryman. Arensberg, Conrad. New York: American Museum Science Books. The Natural History Press, 1968. (paperback). An excellent analysis of rural Ireland by the distinguished Columbia University Anthropologist.

Rambles in Erin. Bulfin, William. Dublin: The Sackville Library, Gill & Macmillan, 1979. A beautiful reprint of William Bulfin's bicycle journey around Ireland. Originally a series of newspaper articles published in book form in 1907; fascinating Victorian adventure.

"John Tyndall—Contributions to the Development of Infrared and Solid State Communications." Callahan, P.S. In *John Tyndall, Essay on a Natural Philosopher.* ed. W.H. Brock, N.D. McMillan and R.C. Mollan. Dublin: Royal Dublin Society, 1981. An essay on John Tyndall's contributions to modern communications.

A View of the Irish. Cleeve, Brian. London: Buchan & Enright, 1983. A book about idiosyncracies of the Irish that is at times a satire and times a recounting of their real character, by a modern TV writer.

Francis Ledwidge, A Life of the Poet (1887-1917). Curtayne, Alice. London: Martin Brian & O'Keeffe, 1972. Probably the most beautiful and readable biography of a poet ever written.

Francis Ledwidge—Complete Poems. Curtayne, Alice. London: Martin Brian & O'Keeffe, 1974. The best ever collection of the poet's poems from his few works.

Gentle Places & Simple Things, Irish Customs & Beliefs. Danaher, Kevin. Dublin: The Mercier Press, 197 (reprinted paperback). A fascinating account of the beliefs of rural Ireland up to the 1940s when they began to lose their magic.

Irish Folk Ways. Evans, E. Estyne. London: Routledge & Regan Paul, 1957. A more complete coverage of Irish country life than Evan's *Irish Heritage.* Its description of booleying and the Irish Cottage are excellent.

Collected Poems. Kavanagh, Patric. London: Martin Brian & O'-Keeffe, 1964 (1972 edition). The best collection of the poet's works with his an "authors notes."

Patric Kavanagh Country. Kavanagh, Peter. The Curragh, Ireland: The Goldsmith Press, 1978. A delightful biography (pocket sized) of the poet by his brother Peter.

Wild Sports of the West. Maxwell, William H. East Ardsley Wakefield Yorkshire, England: E.P. Publishing Ltd., 1973. A reprint of a great 1832 Irish classic about hunting in the west of Ireland. It is far more than a hunting treatise as it is loaded with insights about the life of nineteenth century Ireland in the west.

The Irish in Love. McCann, Sean. Dublin: The Talbot Press, 1972. Everything one would wish to know about the love and courting customs of the Irish countryside.

Life in Donegal (1856-1900). McCarron, Edward. Dublin: The Mercer Press, 1981 (paperback). One of the best descriptions of Donegal country life by a lighthouse keeper. Life had little changed around Belleek by the time of my sojourn in that lovely village.

In Search of Ireland. Morton, H.V. London: Methuen & Co., 1930 (reprinted 1970). A literary masterpiece by the greatest travel writer of all times. His "In Search of England" is equally as good.

The History of Landlordism in Donegal. O'Falleobair, Prionnriar. Ballyshannon: Donegal Democrat Ltd., 1962. A good history of the land war between landlord and the Irish peasantry. The same misunderstanding has moved to the cities these days.

Harvest Home, the Last Sheath. Paterson, T.G. Dundalk: Dundalgan Press, 1975. A marvelous account by a north Irish Protestant businessman-scholar of rural life in Northern Ireland. Proof positive that the hard working gentle north Irishman should be left alone to join or not join the south as they see fit.

IRELAND'S MAGIC

The Stars and the Stones, Ancient Art and Astronomy in Ireland. Brennan, Martin. London: Thames & Hudson, 1983. The best ever account of a considerable amount of research into the astronomical function and artistic meaning of the stone structures of Ireland.

The Soul of the Ghost Moth. Callahan, P.S. Old Greenwich, Conn: The Devin-Adair Co., 1981. A short account of my life and theories of insect communications.

Haunted Ireland, Her Romantic and Mysterious Ghosts. Dunne, John J. Belfast: The Appletree Press, 1977. The best listing of old Irish ghost stories. Beautiful early photographs of Ireland.

Yeats and Magic. Flannery, M.C. Gerrards Cross, Buckinghamshire: Colin Smythe Ltd., 1977. A literary analysis of Yeats' belief in magic as related to his many beautiful poems.

Voices from Stones. Haugaard, Myrna and Brian Lalor. Dublin: The Academy Press, 1983. Beautiful poems by Myrna Haugaard and drawing by Brian Lalor that gives a feeling for the magic of Ireland—so I include them under Ireland's Magic.

The Sphinx and the Megaliths. Ivimy, John. London: Abacus, 1976. A fascinating account of Mr. Ivimy's theory that the stone circles of Britton and Ireland were built by an Egyptian Colony as astronomical observatories for studying the motions of the sun.

The Holy Wells of Ireland. Logan, Patric. Gerrards Cross, Buckinghamshire: Colin Smythe, 1980. An excellent account of the history and curative powers of the Holy Wells of Ireland.

Irish County Cures. Logan, Patric. Belfast: The Appletree Press Ltd., 1981. The best and only book on Irish country cures listed by body organ, mouth, heart, etc., and type burns, sprains, etc.

Irish Earth Folk. MacManus, Diarmuid. Old Greenwich, Conn.: The Devin-Adair Co., 1959. A marvelous account of the Irish traditional beliefs in fairies, Pooka's, Spirits and magic cures. Chapter 10 is an excellent account of Biddy Early.

Irish Wonders. McAnally, D.R., Jr. New York: Weathervane Books, 1938. A reprint of a classic about ghosts, leprechauns, fairies and other magic happenings in the Emerald Isle.

Ireland, A Journey into Lost Time. O'Siochain A.P. Dublin: Foilsiuchain Eireann, 1981. A marvelous account of ancient stone structure that links them to an unknown lost civilization that disappeared about 3300 B.C. with the event of a disastrous comet.

Irish Pilgrimages. Pochin, Mould, Daphne D.C. Old Greenwich, Conn.: The Devin-Adair Co., 1957. An excellent account of history, traditions and customs of Irish pilgrimages.

NATURAL HISTORY

Wild Flowers. Collins Gem Guides. Plamey, Marjorie & Richard Flitter. London: Collins, 1980. An excellent little vest pocket guide of wild flowers of England and Ireland—a must for a Burren walk.

Durman, Roger. Bird Observatories in Britain & Ireland. Berkhamsted: T. & A.D. Poyser, 1976. Excellent chapters on Copland (N. Ireland) and Cape Clear Island (S. Island) bird observatories, directions on how to visit these two island bird preserves.

Ireland's Countryside. Fitzpatrick, H.M. Dublin: David P. Luke, 1973. An excellent description of Ireland's rural countryside and wildlife for young people.

The Birds of Dublin and Wicklow. Hutchinson, Clive. ed. Dublin: The Irish Wildbird Conservancy, 1975. A short ecology and listing of the birds found around Dublin.

Falcons Fly in Ireland. Jocher, Ernest C.F. Cork: The Mercier Press, 1967. An excellent little paperback account of falconry in Ireland between the 11th and 15th centuries.

The Birds of Wexford. Merne, Oscar J. Bord Failte, The Irish Tourist Board, 1974. An excellent little guide to Wexford County, especially of shore birds on Wexford Slobs (coastal reclaimed lowlands). Not a listing but a where-to-find book.

A Natural History of Ireland. Moriarty, Christopher. Cork: The Mercier Press, no date. A generalized paperback guide of the natural history. The first part is ecological principles, the second county by county guide—a good overview.

A Guide to Irish Birds. Moriarty, Christopher. Cork: The Mercier Press, 1967. A generalized family and species listing of birds of Ireland.

The Fauna of Ireland. O'Rourke, Fergus J. Cork: The Mercier Press, 1970. A generalized coverage of the amphibia, reptiles, birds and mammals of Ireland.

Breeding Birds of Britain & Ireland. Parslow, John. Berkhamsted. T. & A.D. Poyster, 1973. An excellent coverage of the breeding birds of Ireland (and England) by family, with the current status and distribution maps.

The Fowler in Ireland. Payne-Gallwey, Sir Ralph. London: John Van Voorst, Paternoster Row, 1932. A fascinating account of the slaughter of wildlife, eagles included, in nineteenth century Ireland.

The Birds and Flowers of the Salter Islands. Perry, R.W. & S.W. Warburton. Privately published, Belfast: Belfast Litho Printers, 1976. An excellent and beautifully illustrated guide to one of the best sea bird islands in Ireland.

The Way That I Went. Praeger, Robert Lloyd. Dublin: Allen Piggis, 1969. A reprint of the greatest classic writing on the natural history of Ireland by the north Irish botanist Robert Lloyd Praeger—a must for any serious student of Ireland.

Natural History of Ireland. Praeger, Robert Lloyd. London: E.P. Publishing Ltd., 1972. A republished version of the 1950 Botanist Praeger's habitat description of Ireland's natural history. An excellent overall view and a must for the serious scholar of the ecology of Ireland. Excellent description of the geology of Ireland from the Paleozoic to the Cretaceous period and later ice age.

Birds of Clare and Limerick. Statleton, L. ed. Limerick: Irish Wildbird Conservancy, 1975. A list of birds of North Munster, including the Burren.

The Irish Donkey. Swinfen, Averil. London: J.A. Allen, 1975. The best account ever of the place of the Irish donkey in rural Ireland by a genuine donkey lover like myself.

Birds of Galway and Mayo. Whilde, Tony. Galway: The Irish Wildbird Conservancy, 1977. A listing of the birds of Galway and Mayo.

SOIL

Soil Physics. Baver, L.D., W.H. Gardner & W.R. Gardner. New York: John Wiley & Sons, 1972. Everything you ever wanted to know about soil physics. No paramagnetism mentioned.

Physical and Geotechnical Properties of Soils. Bowles, Joseph E. New York: McGraw-Hill Book Co., 1979. An excellent book on the properties and formation of soil from minerals and rocks. No paramagnetism mentioned.

Soils in Construction. Schroeder, W.L. New York: John Wiley & Sons, 1980. Paramagnetism not mentioned.

INDEX

INDEX

acceptor hawk, 68
Adcock radio range, xviii
adobe, 185
African grey parrot, 177
age of marine invertebrates, 47,48
Agent Orange, xvii
agriculture, high energy, 132, 133; low energy, 132; stone age, 116
AIDS, 149
Alaska, 49
Albany, 25
alchemy, 121
alcohol molecules, 139
alder trees, 174
All Nature is my Bride, 71
Allingham, William, 166
American Association for the Advancement of Science, x
American Eighth Division, 1
American Monarch butterfly, 34
American yellow-billed cuckoo, 68
ammonia, 139, 143
amnesia, computer, 116
ancestors, Irish, 69; Spanish, 69
Ancient Mysteries, Modern Visions, xiii, 22, 111
ancient stone monuments, 95
aniseed, xv
Anm Maile Inbir Maci Brocann (Mael Inbir son of Brocann), 113
antenna, 120-124, 141, 149, 194; dielectric, 148; ground, 194; insect, 144, 145; loop, 147; magnetic, 194; metal, 144; plant loop, 148; radar, 186; round tower, 196; soil, 194
anti-submarine pilots, xviii
Antiquities of the Irish Countryside, 111
Apache Indians, 24
Apache wars, 195
Appalachian Mountains, 54
arable crops, 92

Aran Islands, 151, 189, 191
Arctic Circle, xix
Arctic stoat, 100
Ardmore, 168
Aristotle, 143
Armorican upheaval, 54
Army Air Corps, xviii, 1, 3, 12, 20, 24
Army Airways Communication System (AACS), xix
arrows, 94
The Art of Mountain Tramping, 31
ash tree, 90
Asheville, xix
Asia, 111, 176
Atha Fada, 109
Atlantic Ocean, 37, 45, 77
atomic bomb, 94
atoms, inorganic, 183
attacks, wolf, 96
attar of roses, xv
Augustinian priory, 156
aura, 75; of rocks, 179
Australia, 37

Ballintra, 43
Ballyliffin, 65
Ballyshannon, 12, 13, 39 40, 43, 86, 87, 166, 168
Baltinglass, 115
barn owls, 104
Barnes' cottage, 6, 42
barracks, 6, 7, 13
barrier, stone-hedge, 94; wind erosion, 99
Barrow, the, 136
Barrow, Professor Lennox, 156, 171
basalt, volcanic, 190, 195
Basho, 153
Battle of the Bulge, 19, 68
Baum, L. Frank, 21, 22
B.C. America, 56

Carlton Hotel, 43
carnac, 38, 39
Carrick Hill, 39
carrick, 5, 39
Carrickfergus, 39
Carrickoris, 39
Carrowheel, 111; mountains, 108, 109
castles, 106
Catalina, flying boat, 57; RAF aircraft, 81
cathedrals, Gothic, 115
Catholics, 178
cats, 104
Cecropia moth, 122
Celtic cross, 177
Cenozoic, 47, 48
cerddin, 88
chakras, 177, 179
chemotherapy, 149
cherry, sour, 90
China, 18
Christ, 68, 151, 196
Christianity, 17
church, Kilmalkedar, 113
circles, stone, 114, 123
civilization, Egyptian, 66, 114, 116
classical physics, 120
Cleary's Hotel, 4, 7, 10, 37; kitchen, 12
Cleary, Amenda, 9, 12
Cleary, baby Mary, 9
Cleary, Des, 9
Cleary, Jimmy, 9
Cleary, Mary, 9
cloves, oil of, xv
coherent radiation, 145; frequencies, 143; lines, infrared, 138; low energy, 145; scatter frequencies, 145
Cole, G. A. J., 54
collected energy, extra, 121
collector, energy, 120, 121, 124
collie dog, 104
Colorado, 22
Committee for the Scientific Investigation of Claims of the

Paranormal, 118
Communication System, xii
Company Donegal, 104
Complete Poems, 172
computer amnesia, 116
conducting loop, 147
conductor, plant, 147
conglomerates, 52
Connemara, 189
Connors, Bridget Ellen, 127
continental drift, 48, 49
control, insect, 104; rodent, 104; electromagnetic, xvii
Conwill, Master, 135
corn earworm moth, 138, 141, 143
corporate feudalism, 87
cosmos, 120, 121, 195
Costellano, Rocky, 6
cottage(s), hip roofed, 78, 80; Irish, 192; magic, 75; stone gabled, 80, 82
countryside, Navan-Slane, 164
County, Carlow, 135; Clare, 62, 125, 162, 170, 189; Cork, 110; Donegal, 3, 65, 132, 166; Galway, 33, 34, 84; Kerry, 55, 113, 131; Meath, 164; Offaly, 39; Wicklow, 115
crannogs, 111
Crataegusmonogyna, 90
creation, viii
Creator, 70
creature magic, 70, 71, 73-75, 88
Creevy, 15
croft, highland, 88
crop(s), arable, 92; yield, 102
cross, Celtic, 177
cross, reproductive, 152; -fertilizing benefits, x
cuckoo, American yellow-billed, 68
cultures, stone, 116
cure, donkey, 131-133, 135, 145, 148, 149, 196; turnip, 131
Curlew mountains, 109
currents, electric, 147; low, 147
Curtayne, Alice, 172

glacier, 45
Glandore, 110
Glenade, 45, 54, 67
Glencar, 45, 54
Glendalough, 154; Round Tower, 53, 154
glens, 106
God, viii, 37, 40, 71, 74, 75, 124, 151, 153, 157, 159, 160, 162, 196, 197
Gorgarty, Oliver St. John, 27, 28
Gormley, John, 88
Gort, 33, 34
Gothic cathedrals, 115
Grand Canal, 39
Grassano, Rocky, 6
Grassmore, 31
gravel, volcanic, 195
Great Britain, 165
Great Gable, 31
Great Pyramid, 96, 119
great tit, black-bibbed, 73, 74, 85
Green Tortrix, 39
Grey, Stephen, 145, 147
Griesdale Pike, 31
grit, volcanic, 64
ground antenna, 194
Gurteenrreagh, 129

Haliburton, Richard, 18
Hall, Richard W., 31
Harmon, Maurice, 34
Harvard University, 56, 67
hawk, accepter, 68; sparrow, 68
haws, 89
hawthorn tree, 89, 90, 92
Hawthorn, Nathaniel, 176, 193
hay, 77, 92
hazel, rods, 92; tree, 90
healer, 126, 129, 131, 132, 147, 196
Heat as a Mode of Motion, xv, 141
heat seeking missles, 94
hedge(s), 102, 104; school, 135; stock-proof, 99
hedgehog, 100
hedgerow, 86, 88, 90, 92, 94, 96,

98-100, 104, 105, 125, 189, 192; rock mound, 92, 93
Heliothiszea, 138, 141
hemlock, 59
herbalist, 129
herb(s), homeopathic, 196; robert, 90
Herschel, Sir William, 141
high energy, agriculture, 132, 133; radiation, 134; systems, 133
high frequency radio, 133
high Sierras, 177
highland croft, 88
hip roofed cottages, 78, 80
Hitler, Adolph, 55
hogweed, 90
Holland, 63
homeopathic herbs, 196
honeybees, xiv
Hooper, M.D., 104
Horner, Michael, 157
Howth Harbor, 25, 28, 29
Hueco Tanks, 22, 24, 69
Hugo, Victor, 28
huts, wattle, 94
hydrodynamics principle, 124

I.R.A., 7
incoherent light, infrared, 143
India, 118
Indians, 70, 197; Apache, 24; Navajo, 24
infrared, coherent lines, 138; incoherent light, 143; lines, 139; low energy, 134; radiation, xv, xvi, 134, 138, 141; radiations, coherent, 145; spectrophotometer, 116; spectrophotometry, xv; wave, 141
inherited, empathy, 74; memory, 24, 25, 34, 36, 66, 67, 69, 159, 160
inorganic atoms, 183
insect, antenna, 144, 145; control, 104; vibrating antenna, 143
insectivorous birds, 99

transmitter, IR, 148; low frequency, xviii; radio, 186
tree, alder, 174; ash, 90; beech, 90, 174; deciduous, 90; elm, 90; hawthorn, 89, 90, 92; hazel, 90; rowan, 88, 192; sycamore, 90; whitethorn, 93, 149
trichomes, 123
Trichplusia ni, 143
Trinity, 178
Tuning In To Nature, xv
turf, 80, 81; Irish, 120; slane, 80, 84 84; spade, 83
turnip cure, 131
Twain, Mark, 21, 183
Tyndall, John, xv, 135, 137, 141, 143, 150, 162, 179, 180, 183
typhoid, 127

U.S. Department of Defense, 135
Ulaidh, 43, 45, 55
Ulster, 51
ultraviolet light, 174
underwater volcanoes, 49
uranium, 153
USDA Insect Laboratory, 67
UV region, 134

vibrating antenna, insect, 143
Vikings, xiii, 167, 168, 170, 171
village system, 64
visible, light, 143; radiation, 134
volcanic, basalt, 190, 195; dirt, 191; fault, 51; gravel, 195; grit, 64; paramagnetism, 194; rock, 180, 183, 189
volcanoes, underwater, 49

Wales, 52, 90, 114
Walls of Limerick, 14
Walls of Venice, 14
Washington, 120
water evaporation, 102, 104, 194
water ouzel, 124

water spider, Giant Nebula, 164
wattle huts, 94
wave, infrared, 141; light, 141; radio, 141
waveguide(s),148; dielectric, 144
wavelengths, IR, 148
Way of the Shaman, The, 157
Webster's Third New International Dictionary, 120
Webster's Unabridged Dictionary, 20
wedge-tomb, 65
Wegener, Alfred, 48
weirs, eel, 123
White dipper, 106
white throat, 100
white water force, 123
White, Gilbert, 100
White, William M., 71
whitethorn, 89-91, 93, 103, 105, 125, 149, 174, 192
whithred, 100, 104
whitterick, 100
Wicklow, mountains, 51; volcanic fault, 52
wild rose, 90
Wiltshire, 111
wind, chill factor, 78; erosion barrier, 99; erosion control, 104; speed, 102; -blown erosion, 104, 105, 191
Winding Banks of Erne, The, 168
witch of County Clare, 126, 150
witchcraft, 126
wolf attacks, 96
wolf packs, 96
wood, dead, 100
Wordsworth, William, 153
working society, stone, 114
World War I, xx
World War II, ix, xviii, 6, 26, 28, 43, 58, 104, 152, 153, 156, 188
worship, megalithic, 115
Wright, John Cardinal, 192
Wuthering Heights, 59

x-ray region, 148